Aerodynamic Interference and Rotor Optimization in Multirotor UAVs

Tiyagi

CHAPTER - 1

INTRODUCTION

1.1 INTRODUCTION TO UAVS

Drones have a wide range of applications across various industries. It provides exceptional performance in agriculture, environmental protection, infrastructure inspection, construction and rescue, Security, Surveillance, Scientific Research, Personal, Recreational Use and Delivery Services. The first known use of a drone dates back to the Austro-Italian War in 1849. Drones continued to advance during the Cold War and the War on Terror, becoming more and more important in military operations and surveillance. Drones have been used on an unprecedented scale in the Ukrainian war, evolving from specialized instruments to vital weapons incorporated into the organizational frameworks of both forces. Since their early inception, drones have played a crucial role in both modern aviation and warfare. Apart from these other sectors, it also has become integral to logistics, and supply chain management (SCM). It is transforming the way how the goods are transported and managed.

However, the overall efficiency and quality of transportation is still disturbed by the aerodynamic and stability influence of rotors. Being the barrier for the future scope of drones, the interference effects and other challenges have to be addressed and revolutionized to reshape the payload transportation. Another interesting grouping of several unmanned aerial vehicles is called a drone swarm. These vehicles must operate in close proximity to one another, communicate, synchronize their motions, and sometimes even operate autonomously to accomplish predetermined goals. They function as unified and coordinated entity where maintaining the closer operating distance plays a significant role to achieve the desired task.

1.1.1 Classification Based on Configuration

a) Fixed-Wing UAV

Fixed-wing UAVs function without a pilot on board, resembling conventional aircraft. Their primary purpose is to perform specific missions such as pesticide spraying, surveillance, and reconnaissance. Missions requiring lengthy endurance and long range are usually better suited for fixed-wing UAVs.

b) Rotary-Wing UAV

The rotorcraft, another name for the rotary-wing UAV, is very different from the other structures. It produces the lift required for flight with its revolving wings, which are fixed within the vehicle's structure. Medium-range tasks usually suit these rotorcraft UAVs. Two types of rotary-wing UAVs can be distinguished further:

c) Single-Rotor Vehicles

A second vertical rotor assists in steering the single-rotor UAV in the desired direction while the main rotor produces the lift required for flight. One notable instance of this kind of aircraft is the Single-Rotor Helicopter. However, there is a lot of complexity involved in operating the single-rotor helicopter. It is more sophisticated than multirotor UAVs, but it has advantages of its own. However, the Single-Rotor Helicopter's control system is highly intricate. Notwithstanding its complexity in comparison to multirotor UAVs, it has advantages of its own.

d) Multi-Rotor Vehicles

Among other things, multirotor include quadrotors, octocopters, and more. The adaptability and breadth of capabilities that drones offer to diverse industries and applications are demonstrated by each of these setups.

e) Hybrid UAV

The hybrid-wing unmanned aerial vehicle (UAV) blends characteristics from two different classes into one: the fixed-wing and rotary-wing UAVs.

1.1.2 Classification Based on Size

a) Very Small UAV

Very small UAVs are between 30 and 50 centimetres long, or the size of a giant insect. UAVs that resemble insects and have rotating or flapping wings are a common micro design. They can be utilized for biological warfare and espionage because of their small size and low weight. Bigger ones use standard airplane configuration. The desired level of manoeuvrability determines whether to use rotating or flapping wings. Designs with flapping wings enable perching and landing on tiny areas. The Israeli IAI Malat Mosquito, which has a 35 cm wing span and an endurance of 40-minute, is one example of a very small UAV.

b) Small UAV

Small UAVs, commonly known as mini-UAVs, are classified as UAVs with a minimum dimension of more than 50 cm and a maximum size of no more than two meters. Most of the designs in this category are hand-launched by tossing them into the air, and many of them are based on the fixed-wing form.

c) Medium UAVs

UAVs that weigh too much for a single person to carry but are still smaller than light aircraft are classified as medium UAVs. They can carry cargoes weighing between 100 and 200 kg, and their typical wingspan is 5 to 10 m. Figure 1.6, below, shows two examples of medium fixed-wing UAVs:

the Israeli-US Hunter and the UK Watchkeeper. Other names that have been used in the past include the RQ-2 Pioneer, the US Boeing Eagle Eye, the BAE systems SkyEye R4E, and the RQ-5A Hunter. The Hunter's long is 6.9 m and has a 10.2 m wingspan. About 885 kilograms is its take-off weight.

d) Large UAVs

The large UAV class covers the huge UAVs that the military primarily uses for combat missions.

1.1.3 Classification Based on Range and Endurance

a) Close-range UAVs and Very low-cost

UAVs in this category have an operating range of 5 km, an endurance duration of 20 to 45 minutes, and an approximate cost of $10,000 (estimated in 2012). The Raven and Dragon Eye are two UAVs of this type. This class of UAVs is extremely similar to airplanes model.

b) Close range UAVs

UAVs of this class have an endurance time of one to six hours and a range of fifty kilometres. Usually, they are employed in observation and reconnaissance missions.

c) Short range UAVs

UAVs range of at least 150 km and endurance periods of eight to twelve hours fall into this class. They are mostly used for reconnaissance and surveillance, much like the close-quarters UAV.

1.1.4 Excellence of Multirotor UAV

As multi-rotor UAVs combine affordability, manoeuvrability, and versatility, they are utilized for a plenty of useful applications in the drone sector. Fixed-pitch blades are used in multi-rotors to control attitude and produce thrust. Every rotor is usually driven by a direct-drive electric motor, which results in a mechanical system with relatively few moving parts. Usually, the pitch and blade profile are chosen to maximize performance within the primary flight envelope. Using a technique known as differential thrust, multi-rotors regulate the angular velocity of each rotor independently as opposed to actively changing the pitch of individual blades. All rotors are tuned to the same angular velocity, which produces zero net torque and ensures continuous lift. By varying the angular velocities of the rotors on opposite sides and producing a moment about the centre of mass, rotation in pitch and roll is produced. It is possible to adjust the net torque of the system in the yaw axis by means of counterrotating rotor pairs. Zero torque is applied in the yaw axis when the two rotors' angular velocities are identical. A yaw moment will occur if these rotors' speeds are changed differently. It is necessary for a MUAS to have at least four rotors with pairs of clockwise (CW) and counter-clockwise (CCW) rotors in order for it to be able to operate only on differential thrust in six degrees of freedom. As a result, quadrotor systems which have four rotors have emerged as the most widely utilized multi-rotor configuration in both commercial and research applications.

Multi-rotor propulsion systems' intrinsic mechanical simplicity makes it possible to design airframes in a variety of ways to meet mission settings, payload locations, and performance requirements. Though there are theoretically infinite configuration options, real-world applications and constraints on size, weight, and complexity have produced a set of standard configurations that are employed in both industry and research. A multi-rotor's overall rotor count is determined by a number of factors. The highest thrust that may be obtained from a multirotor of a certain size may be limited by the number of propulsion system components. The maximum available thrust of the propulsion system can be increased by increasing the number of parts that have a fixed size. As the number of rotors rises, redundancy usually becomes more effective. The drawback of adding more rotors is that it makes the system larger, more complex, and heavier. More motors, electrical parts, and mounting hardware all contribute to more failure spots. Because of the increased disk loading on each rotor and the increased number of moving parts, employing more smaller rotors within a given airframe size may also result in a lower efficiency than using fewer bigger rotors. Size constraints, the number of rotors employed, payload mounting locations, and requirements for the attitude control authority can all influence the rotor mounting configurations. Rotor configurations that are frequently used are thrust vectoring, overlapping, coaxial, and radial.

In comparison to other kinds, multi-rotor UAVs are the least expensive choice. The most efficient uses of multi-rotor UAVs are aerial surveillance and photography because they offer the most control over location, framing, and mobility. These drones have the ability to move vertically up and down in tandem with front-to-back motions, spin on their own axis, fly near to buildings, and move side-to-side. Based on the number of rotors incorporated in their design, multi-rotor UAVs are divided into four subclasses: tricopters, quadcopters, hexacopters, and octocopters.

- Tricopters: These are comparatively simple machines with three rotors.
- Quadcopters: The most common kind of multirotor aircraft are quadcopters, which have four rotors arranged in an X or square pattern.
- Hexacopters: With six rotors, they can carry heavier loads and are more stable.
- Octocopters: These are utilized for specialized and heavy lifting tasks, and they have eight rotors.

1.1.5 Introduction to Quadcopters

An unmanned aerial vehicle (UAV) or drone with four rotors each with an engine and propeller is called a quadcopter. It is sometimes referred to as a quadrotor or helicopter. Quadcopters are employed in many different fields, including research, surveillance, aerial photography, and filming. The military also uses them for combat and reconnaissance missions. The small quadrotor's viability in this application has been substantially enhanced by its small size and low inertia, which allow for the employment of a very straightforward flight control system. Quadcopters, in contrast to traditional helicopters, typically lack cyclic pitch control, which allows the blades to change angle dynamically as they rotate around the rotor hub. Rather, separate changes in each rotor's speed, and consequently its lift and torque, enable flight control. Changes in the net torque are what regulate yaw, while adjustments to the net centre of thrust control pitch and roll. Generally, a quadcopter consists of two clockwise (CW) and two counter-clockwise (CCW) rotors.

1.1.6 Applications of Quadcopters

Drones can be remotely controlled, which provide amazing vistas from above and are excellent for taking aerial pictures and films. They're also useful for delivering packages, particularly to locations that are difficult to access. These are only a handful of the numerous applications for drones that demonstrate their great value. In addition, farmers are also using drones to survey their crops from the air. Furthermore, drones are utilized in search and rescue operations to locate individuals in difficult-to-reach locations. Drone usage is revolutionizing emergency response and agricultural.

1) Agricultural Application

Significant use of technology is in agriculture, where drones are providing farmers with efficient and reasonably priced ways to monitor their crops. Infrared sensor-equipped drones are particularly useful for crop health monitoring, providing farmers with quicker and more accurate information to take preventative action to enhance crop conditions and production. In the next several years, drone technology is expected to impact 80% of the agriculture sector.

2) Terrain Mapping

Drones equipped with 3D cameras and LIDAR sensors are being used by scientists and researchers to map topography in three dimensions. The AI-powered drones don't require human assistance to fly over a predetermined course because they are preprogrammed to do so. The drone has the ability to transmit live films and photos to ground stations for processing, or it can save recordings onboard. In particular, this drone application is helpful for producing strong and accurate Digital Elevation Models (DEM) of terrain, which are utilized in many scientific fields such as GIS and remote sensing.

3) Livestock Management

Drones are utilized not just for agriculture management but also for animal management. In order to improve livestock management and tracking, ranchers can pre-program the AI-powered drones to follow and monitor cattle in the fields. This allows the ranchers to see the livestock from the air in real time. Drones with thermal sensor mounted are also employed for real-time problem identification and overnight livestock surveillance.

4) Solar Panel Inspection

With its explosive expansion and development in recent years, solar energy has emerged as a significant renewable energy source. Massive solar energy parks have been built by nations all over the world to produce clean, renewable electricity. Authorities may use dependable, consistent, and reasonably priced technology to evaluate solar panels and find any possible problems that could reduce the effectiveness of the solar parks by using an AI drone solar panel inspection system that is outfitted with several sensors and image detectors. Not only can AI drones provide a trustworthy and excellent way to monitor solar parks, but they are also far more cost-effective and dependable than human inspection.

5) Aerial Photography

One of the initial uses of drones was for aerial photography. During World War I, a number of nations employed air balloons, the ancestor of drones, as reconnaissance devices to learn about the whereabouts of their enemies. Drones are now being employed as a far less expensive option to helicopters for filming footage of isolated and unreachable locations. Drones are employed all around the world as the most effective, adaptable, and economical instrument for taking beautiful aerial photos, by everyone from individual Vloggers to major Hollywood production companies. Drones are even being used by news organizations to get first-hand data and transmit live.

6) Shipping and Delivery

Even while drone technology is still relatively new, some of the largest names in the digital world, like Amazon and DHL, are embracing shipping and delivery. By flying to the sky, delivery companies may improve customer service and expedite their delivery system. In particular, Amazon is advancing drone technology by using an AI positioning system to provide same-day package delivery to customers.

7) Geographic Locations

Drone technology is being used more and more by professional and government agencies to obtain high-definition imagery of isolated areas such as rocky slopes, inaccessible mountaintops, islands, and coastlines. Contemporary drones are equipped with a variety of imaging sensors that can be used to crowdsource mapping projects or to produce three-dimensional maps of geographical areas.

8) Geographic Locations

Drone technology is being used more and more by professional and government agencies to obtain high-definition imagery of isolated areas such as rocky slopes, inaccessible mountaintops, islands, and coastlines. Contemporary drones are equipped with a variety of imaging sensors that can be used to crowdsource mapping projects or to produce three-dimensional maps of geographical areas.

9) Disaster Management

One of the most significant and extensively utilized technological technologies that has been embraced globally is disaster management. Drones are the most effective and economical instrument for gathering data from ground zero and creating relief operations plans because of their special capacity to reach disaster-stricken areas more quickly and safely. Drones equipped with specialized sensors, cameras, and radars can help rescue personnel develop preparations for relief and rescue efforts by providing a comprehensive picture of the extent of the crisis.

10) Search and Rescue

Drones can now be effective instruments for surveillance at night thanks to the thermal sensors installed on them. Drones using infrared and thermal sensors can be equally effective day and night when tracking missing people through forests, mountains, and other difficult terrains. Drones of today can function well in challenging environments as search and rescue missions.

Drones are useful not only for searching for missing people but also for airdropping goods to areas devastated by conflict or natural disasters. Drones, for example, can be equipped with a GPS tracker, food, medicine, and other necessities to help stranded individuals live until rescue arrives.

11) Weather Forecast

Creating dedicated drones for weather forecasting and monitoring is a new trend in drone technology. Drones are being used by weather and climate agencies all over the world to track and monitor dangerous weather patterns, such as hurricanes and low-pressure systems over oceans, in order to gather firsthand information about the extent, behaviour, and trajectory of the event and share it with the local population.

L2) Wildlife Monitoring

Drones have long been used in Africa as a means of discouraging poaching. Drones are an ideal technological advancement for monitoring wildlife in deep forests or vast grasslands because of their exceptional capacity to maintain surveillance over big areas continuously in all weather conditions. Drones are not only employed by scientists to track endangered species and research their behaviour and habitat, but they also serve as a deterrence against poachers.

13) Law Enforcement

Drones are extensively used by law enforcement organizations to uphold law and order. Drone application becomes even more important during protests, as technology allows law enforcement to regularly and closely monitor the protest scenario and identify any unlawful event before it gets out of control. Drones are employed not only to keep the peace in urban areas but also by border control organizations to track drug traffickers and their illegal goods as they cross coastlines or rural, unreachable locations.

14) Entertainment

In the entertainment business, drones are being used more and more for fight clubs, concerts, and even just filming material for videos. AI-powered drones give singers and vloggers the extraordinary capacity to record the breath-taking beauty of the nature or create entertaining videos from unusual angles. Drones have also been utilized more recently in fight clubs, where two drones engage in combat alongside their opponents.

15) Military Reconnaissance Uses

Drone technology was first employed by the military to collect vital enemy intelligence from the battlefield. Today's technology has only increased its effectiveness in obtaining vital information, not just from the battlefield but also in times of peace. These days, formidable drones outfitted with the most potent imaging technology and soaring thousands of kilometres above the sky are able to compile comprehensive data regarding the enemy's armed forces within its borders. Additionally, border surveillance using small quadcopters is a common practice to obtain real-time intelligence on adversary forward positions.

16) Military Combat Uses

In addition to its usage in reconnaissance missions, the technology is employed in direct combat situations. The largest producer and user of military drones, the United States, has been utilizing the technology for decades to battle in harsh and dangerous environments, protecting its soldiers while taking out enemy objectives from a distance.

1.2 PAYLOAD TRANSFER USING DRONES

Drones are perfect for logistics and transportation since they can carry and deliver items thanks to drone payload delivery systems. Delivery efficiency and speed can be greatly increased with the use of payload release mechanisms, particularly in crowded urban areas. Since drone deliveries provide a quicker, more cost-effective, and environmentally friendly alternative to conventional ground-based transportation methods, businesses like UPS and Amazon are already investigating drone deliveries.

Payloads can be categorized into four groups: non-dispensable, disposable, active, and passive. These groups are based on how the payloads are attached to the drone and how they function during the mission. Deliverable payloads to the end user, disposable payloads can be released from the aerial vehicle at any point during the flight in response to a signal received from the ground control station (GCS) or radio controller (RC). Throughout the operation, non-dispensable payloads stay physically attached to the UAV, yet their contribution is essential to its success. Throughout the operation, Active Payloads are either fully or partially active, serving a variety of purposes such as mapping, data collection, surveillance, and so forth. The majority of passive payloads can be delivered at a predetermined location and are inert during the mission.

a) Payload Transfer in Mid-Air Using Multi-Drones

Recently, collaborative aerial work has drawn attention for a number of reasons, primarily related to long-range coverage but also includes enhanced control, ease of maintenance, enhanced safety, complicated shape transfer potential and greater capacity for payload. Enhancing the durability of payload transportation systems in single and multi-UAV has the potential to expand their coverage area, hence increasing their suitability as a delivery and transportation substitute for traditional methods used.

b) Drone to Drone Payload Transfer

Drone cargo transfers in mid-air are difficult and demanding tasks. Nonetheless, a number of techniques, such as robotic arms, magnetic coupling, and tethering, can be used to move supplies between drones. Even in the event that one drone's charge runs low, these techniques can be utilized to transfer payloads between drones in the air. Using a tether or magnetic coupling, for instance, a drone with a low battery can transfer its payload to another drone, which will then supply the necessary power to maintain the payload in the air. As an alternative, payloads can be transferred between drones in flight by robotic arms. However, only skilled specialists should attempt these approaches as they need for certain tools and knowledge. It is crucial to remember that moving payloads between drones in the air can be risky and needs to be done carefully. It is essential for pilots to possess a comprehensive comprehension of their drones' capabilities and limitations, along with the necessary equipment and tactics for ensuring a safe payload transfer. Pilots may make sure they are utilizing their drones safely and productively by adhering to safety requirements and best practices.

c) Dual UAV Collaborative Payload Transfer

Using this method, two drones collaborate when transporting a payload. Depending on how full their batteries are, the drones divide the load and power usage among themselves. The approach is based on load sharing and energy distribution. In a recent study, two UAVs are used to collaboratively transport a slung load using a real-time dynamic programming (RTDP) technique with a trajectory adjusted for both energy and time usage. The research discusses the application of the RTDP algorithm. The path was split into distance interval horizons, and the optimal strategy for each distance interval was determined using a dynamic programming sweep. The RTDP-based method for dual-UAV cooperative payload transit uses coordinated motion, with UAVs acting as controllers on the payload. The RTDP algorithm finds the best velocity selections for the slung load transfer in order to minimize the use of energy or travel time. The RTDP method involves reducing a cost factor that is derived from splitting down the combined model of 'the dual-UAV-payload system. The cost function calculation was also considered in order to dynamically shift the load/energy across two multi-rotor platforms over a transportation operation. The cost function is used to calculate the transition costs for each stage and velocity decision. A terminal cost is imposed at the final distance gap during the initial part of the journey, when the velocity at the end of the current horizon is uncertain. The final velocity, which is used to calculate the transitioning cost of the final stage, is known since the destination is a part of the last stage, or edge of the horizon, in the next stage. The RTDP-based strategy is acting as actuators on the payload. For the slung load transfer, the RTDP algorithm determines the ideal velocity choices to reduce the energy consumption or the travel duration.

1.2.1 Challenges Faced During Payload Transportation

a) Distance Between Vehicles

A system that permits two vehicles to swap payloads in mid-air needs to extend the range of their nearest approach. Hovering close together increases the chance of a collision for the vehicles because of the unpredictability of the surrounding environment. Consequently, determining the minimum operating distance is required to guarantee the drones' safe interaction.

b) Relative Positioning Accuracy

Relative positioning accuracy is essential for safe and successful UAV payload transfer. GPS information is frequently used by UAVs for positioning. However, there are a number of variables that might impact GPS accuracy, including low satellite visibility, multipath interference, and signal noise.

A method for increasing precision is relative placement. For instance, depending on their distance from one another, a UAV can get corrections from a nearby Unmanned Ground Vehicle (UGV). Inaccuracies in the UAV's GPS data can be made up by using the UGV's data.

In actuality, the UGV can employ several subpar GPS modules as long as they are in line of sight. Based on their estimated distance from one another, the UGV provides GPS data updates to the UAV; the UGV frequently uses a standard image pattern for scaling.

d) Forces Act in Contact of Multiple Vehicles

Knowing the contact forces involved in payload transport with unmanned aerial vehicles is essential. In order to manage oscillations, deal with wind disturbances, and measure cable stresses during take-off, it is imperative to comprehend the force exerted by the payload. not just the forces exerted on the payload, but also the forces that combine when two vehicles contact. Drag will rise as a result of the interaction's disruption of the necessary thrust generation and flow between them, which also disrupts stability.

Operations are safe and stable when the forces acting during payload transfer are understood. It aids in preventing abrupt changes in direction or erratic motions that could damage the load, the surroundings, or the UAV. When the UAV comes into contact with the cargo, precision control is made possible by accurate force measurements. This is necessary to keep things balanced, prevent accidents, and make successful transfers. By measuring forces, it is possible to monitor the payload's behaviour during transfer. Finding abnormalities (such as sudden jerks or strong forces) aids in the early detection of possible problems. The UAV's control algorithms are calibrated with the use of force measurements. Real-world data can be used to inform adjustments, which will enhance performance overall. Force data can be used by researchers to examine various payload kinds, transfer scenarios, and environmental factors. Improved payload transfer system design is influenced by this understanding.

1.2.2 Motivation for Analysing Thrust and Aerodynamic Interference Effects

Despite the fact that MUAS offer mechanically straightforward systems that are adaptable enough to be used in a variety of settings, there are still a number of issues that prevent these systems from being further improved. There are numerous operational and safety standards that must be adhered to while transferring payloads between drones. Transferring payloads between drones can be a challenging procedure. Making sure that the cargo is firmly fixed to the drone and does not fall off during the transfer is one of the biggest concerns and ensuring the cargo is delivered freely and without causing any harm to the drone or the payload presents another difficulty.

In addition to the safety factors, performance parameters have also to be checked to make an optimal and energy efficient design. Improved propulsion system efficiency to provide longer range and endurance or lower the mass and size of required power sources, faster flying speeds and increased manoeuvrability and an expanded payload capacity are all necessary for improved stability in turbulent situations and bad weather. When carrying a payload, UAV thrust management is an extremely important issue to take into account for stability in payload exchange in multi-drones. Due to battery restrictions, multi-rotor UAVs have limited endurance, so conservation of energy is crucial for the aerial transportation of both single and dual-UAV.

The study of the interactions between unmanned aerial vehicle rotors is essential for maximizing its performance in relation to the low energy density of current batteries. Increased inter-rotor wake interactions and decreased rotor efficiency can result from rotor interference. This can affect the UAV's performance. Increasing the number of rotors is also a typical method to improve UAV functionality. On the other hand, as the number of rotors increases, there could be interference from the downwash flow between neighbouring rotors. The interference may cause variations in the UAV's power usage.

Optimizing UAV performance, stability, and efficiency requires an understanding of and mitigation of aerodynamic interference effects. In order to overcome these difficulties in the construction and use of rotor-based UAVs, investigations are still being conducted for creative alternatives.

1.2.3 Interference Effects in Swarm

Interference effects between UAV rotors can significantly impact the behaviour and performance of a swarm of UAVs in several ways:

1. **Coordination and Communication**: Interference between rotors can disrupt communication signals among UAVs within the swarm, making it challenging for them to coordinate their actions effectively. This interference can lead to communication delays or even signal loss, affecting the swarm's ability to maintain formation and perform collaborative tasks.

2. **Collision Avoidance**: Interference effects can distort the sensing capabilities of individual UAVs within the swarm, making it difficult for them to accurately detect and avoid collisions with nearby obstacles or other UAVs. This can increase the risk of mid-air collisions and compromise the safety of the swarm.

3. **Formation Stability**: Interference between rotors can introduce unpredictable disturbances in the airflow, affecting the stability and coherence of formation flight within the swarm. UAVs may experience turbulence or aerodynamic interactions that disrupt their formation geometry, leading to deviations from desired trajectories or formation patterns.

4. **Energy Efficiency**: Interference effects can increase power consumption for UAVs within the swarm, as they may need to compensate for disturbances in airflow or adjust their control inputs to maintain stability. This can reduce the overall energy efficiency of the swarm, leading to shorter flight durations and decreased mission endurance.

5. **Task Performance**: Interference between rotors can degrade the performance of individual UAVs within the swarm, particularly for tasks that require precise control or coordination. Tasks such as payload delivery, surveillance, or mapping may be affected by interference effects, leading to reduced accuracy or efficiency in completing the task objectives.

6. **Swarm Scalability**: Interference effects can limit the scalability of UAV swarms by imposing constraints on the maximum number of UAVs that can operate within a given airspace or environment. As the size of the swarm increases, the cumulative interference between rotors may become more pronounced, limiting the swarm's ability to operate effectively.

To address these challenges, researchers and engineers are exploring various techniques such as distributed control algorithms, adaptive communication protocols, and aerodynamic design optimizations to mitigate interference effects and enhance the performance of UAV swarms in real-world scenarios.

1.3 OBJECTIVES AND SCOPE

With the aim of investigating interference effects between rotors, enhancing thrust, and finding the closer proximity of two operating propellers, the experimental and numerical analysis are carried out for its performance evaluation. The key reason for performing this research is to get an in-depth insight into the formation of wake and its negative impact in drones. The higher the aerodynamic interference, the lower the aerodynamic performance, stability and control precision there will be. Moreover, power consumption, noise level, vibration, wear and tear will also be increased. An attempt is made to find the optimal approaching distance between drones within the safe range and the exploration in this respective distance will be extremely beneficial in payload transfer more safely and swift transfer can be ensured without having added any complex payload transfer systems. Due to the efficient use of battery, it will be useful in long-range payload transportation and continuous monitoring. Understanding and addressing these interference effects are crucial for the successful deployment of UAV swarms, especially in applications requiring precision and reliability.

The following objectives are identified to conduct both the experimental and numerical tests.

- ❖ Design and fabrication of propeller interference test rig with three directional accessibilities.

- ❖ Conduction of experimental tests for the selected 12 inch and 13-inch propellers at different axis swift in both the axial and lateral direction from 4000 rpm to 8000 rpm.
- ❖ Quantification of differences in thrust and figure of merit with respect to spacing and also direction.
- ❖ Numerical simulation to understand the influence of wakes and other flow features with respect to different Axis Swift (AS) and speed.
- ❖ Estimating the minimal closer operational distance between two rotors by the comparative evaluation of experimental and numerical results.
- ❖ Formation of mathematical modelling for generating vehicle path trajectory, altitude, roll and pitch. The vehicle trajectory has to be optimized for the required time and energy consumption before real fly.

1.4 ORGANIZATION OF THESIS

The thesis general organization is as follows:

Chapter 1 It gives a broad overview of UAVs and their superiority in transportation. It also covers its principles, objectives, challenges and significance of examining interference effects as well.

Chapter 2 It covers the studies that have already been done in this field. The experimental and numerical methods followed from various authors are summarized with the obtained results. The chapter starts with quadcopter performance. The interference effects, aerodynamic studies, mathematical modelling and payload transfer between UAVs are presented. It concludes with the problem statement and scope this research identified from the literature survey.

Chapter 3 illustrates the methodology that outlines the research flow and discusses the description and development of the experimental setup. It also presents the specifications and distinguished reasons for the used propeller, motor and thrust stand. The detailed procedure of the experimental tests is also discussed. In addition to it, the numerical design of the propeller. The Mesh generation, CFX setup, and CFD analysis are also explained elaboratively. The design of the model is done in CATIA, meshing and computation are completed in Ansys CFX. The turbulence model used for propeller analysis is K-ε. Mathematical modelling procedure is also discussed elaboratively with the steps to be followed.

Chapter 4 gives a discussion of the results. The experimental, numerical and mathematical results are presented and compared to each other. The quantitative thrust measurement at different axis swift shows the interference effect clearly in between the rotors. The aerodynamic coupling of wakes at different spacing is shown in different contours and the comparative results are also illustrated.

Chapter 5 A condensed summary of the research provides the full details of both the experimental and numerical results on interference effects. The methodology followed is explained clearly and from the comparison of the results, the minimum closer operating distance is found. The percentage of interaction effects with respect to spacing in both directions is presented in this chapter. From the mathematical modelling, the developed trajectory path is explained. The future scope of the work as well as the references are listed.

CHAPTER - 2

LITERATURE SURVEY

2.1 STUDIES ON QUADCOPTER PROPELLER PERFORMANCE

Lei, Y. and Wang, H. (2020) used experiments and computational computations with rotor axis shift from 2.2 to 4.0 to investigate the performance of a quadrotor. The rotor systems' interactions are determined by power, thrust, pressure, the quadrotor's velocity vector, and vorticity. This suggests that altering the rotor spacing actually improves the performance of the quadrotor considerably in aerodynamic point. When compared to a single rotor operating without interaction, the quadrotor systems thrust is evidently enhanced with varying rotor separation. This demonstrates the total hover efficiency of a quadrotor can be increased by a minor degree of aerodynamic interference between neighbouring rotors. At 3.2 and 3.6 L/R, the quadrotor's thrust is higher than that of other cases. The highest thrust increase is approximately 36.8% at the 3.2 L/R. The aerodynamic performance is reduced when the spacing between the rotors is small because of the aerodynamic interaction between the rotos.

The aerodynamic performance of a small quadrotor in the presence of disturbance from horizontal wind was studied by Lei, Y., and Wang, H. (2020). Utilizing a combination of simulations, experiments, and computational fluid dynamics (CFD), the research examines how wind affects the aerodynamic properties of the quadrotor. The findings show that when wind is present, the flow field becomes more complex, resulting in mutual interactions from the rotor wake distortion and the front rotors' downwash flow colliding with the direction of the wind. Because of the increasing pressure differential between the lower and upper surfaces, the thrust of the rotors progressively increases with wind speed.

On the other hand, significant interactions at greater wind speeds may also generate vibration, which increases power consumption. According to the paper's conclusion, at the wind speed of 2.5m/s, the quadrotor exhibits good wind resistance and increased aerodynamic performance. All things considered, the study offers insightful information about how quadrotors behave aerodynamically in windy conditions, emphasizing how wind affects thrust, power usage, and overall aerodynamic performance.

CFD analysis of an 11-inch quad-rotor UAV propeller utilizing an overset mesh technique at 5000 RPM and different airspeeds was presented by Oktay, T., and Eraslan, Y. (2020). With a difference ranging from 14–18%, the numerical estimates for thrust coefficient nearly matched the experimental observations. Furthermore, as predicted, a drop in the thrust coefficient was seen at higher airspeeds. In terms of thrust coefficient, the CFD findings showed trends that were comparable to the experimental values, with the numerical predictions convergence toward the experimental results at lower airspeeds. Overall, the study shows that CFD analysis can accurately forecast UAV propeller performance, especially at lower airspeeds, and offers insightful information for UAV propulsion system design and optimization. Mutual interference causes an 11% thrust loss and a 35% pitching moment loss to the rear rotors at flight speeds of 5 to 15 m/s. Nonetheless, the interference is found to be insignificant in scenarios including hovering and forward flight at 25 m/s. The design and optimization of quadcopter systems can be influenced by these findings, which offer interesting details about the aerodynamic interactions among quadcopters at various flying speeds.

Ye, J., et al., (2021) investigated the mutual interference phenomenon in quadcopters using computational fluid dynamics. The researchers simulated the flow field around the quadcopter at different speeds and introduced "Virtual Modes" for analysing analyse the aerodynamic interactions among its components. These findings reveal that the front rotors' mutual interference is minimal, while the rear rotor's is primarily because of the wake of the fuselage and mainly front rotors. Additionally, interference to the fuselage is produced by the rotors located the front and rear. The study demonstrates that only the pitch and thrust from the rear rotor, as well as the lift of the fuselage, are significantly affected by mutual interference. At flight speeds from 5 to 15 m/s, mutual interference leads to the thrust loss of 11% and a loss of pitching moment is about 35% for the rear rotors. However, in hovering and 25 m/s forward flight scenarios, the interference is found to be negligible. These findings provide valuable insights into the aerodynamic interactions within quadcopters at different flight speeds and can inform the design and optimization of quadcopter systems.

The study by Kuantama, E. and Tarca, R. (2017) investigates how to make a 406mm propeller more efficient by incorporating a cover, or duct, to create a ducted propeller. Ducted propeller types α, β, and γ produce thrusts of 1.38N, 2.1N, and 2.1N, respectively, at a highest angular velocity of 9000 rpm. For types α and γ, the extra ducting frame material shouldn't be more than 140gf and 214gf, respectively. Because type γ has less mass and produces the same thrust, it performs better than type β. In order to avoid propeller interference, ducted propellers are smaller than the ones without ducts, with the body frame size determined by the angular velocity of the propellers. A 460mm squared quadcopter ducted body frame is used for 5000 rpm angular velocity. When comparing power requirements, ducted propellers outperform non-ducted ones because of the airflow resulted pattern produces maximum thrust with the least amount of disturbance.

2.2 STUDIES ON INTERACTION EFFECTS IN UAV

In a 2018 study, Shukla, D. and Komerath, N. investigated how the wakes from two rotor arranged side-by-side rotors interacted depending on their Reynolds number and distance from one another. At low Reynolds numbers, it noticed a drop in performance for cases when the rotors were very close to one another. The efficiency drop may be caused by the interaction between blades of one rotor with the nearby rotor blades tip vortex cores during least axis shifts, according to the researchers' theory, though more testing is required to prove it. Measurements of the flow field revealed strong rotor wake interactions for modest axis spacing, leading to the distortion, loss of coherent shape, and deviation of rotor wake features from their usual paths. The inflow-outflow velocity profiles and time-averaged streamline showed similarities throughout the variety of tests. For the test scenarios, mean velocity profile and significant instantaneous flow characteristics attributes were found. These findings should be about performance estimations and rotor wake across a larger range of flight conditions and vehicle designs.

The rotor-to-rotor interactional aerodynamics in edgewise flight of a quadcopter was the subject of Misiorowski, Gandhi, and Oberai's (2019) study. They especially examined the quadcopter's cross and plus configurations. The tip vortices on the sides of the rotors affected the downstream flow, causing lift and pitching moment change for adjacent rotors. These important findings were made by the researchers using flow field simulations to study a variety of behaviours. Due to interaction effects, the aft rotors in the cross configuration produced less lift and saw a considerable reduction in pitching moment; in the plus configuration, on the other hand, the west and east rotors observed increases in net lift as a result of the north rotor's wake-induced effects. For the quadcopters with the cross and plus configuration, the recorded changes in rotor thrust and moments as a result of the interactional aerodynamic effects were the most substantial. While the plus configuration produced more thrust and used less power in forward flight, the interaction effects caused a significant loss of thrust for some rotors in the cross configuration, needing higher RPM to compensate for the lost thrust.

In their 2020 work, Lei, Y. and Wang, H. combined experimental and numerical methods to discuss the quadrotor's aerodynamic performance with varying the spacing between the rotors in hover. To determine the best configuration for aerodynamic performance, experimental thrust and power consumption measurements were made over a range of rotor spacings (from 2.2 to 4.0). In order to evaluate aerodynamic interference between nearby rotors, computational fluid dynamics (CFD) analysis was utilized to depict the vorticity, velocity vectors and pressure distribution at the blade tip and in the region of flow field. According to the study, there was a detrimental effect on hover efficiency for both excessively small and large rotor spacings, and the quadrotor's performance changed dramatically with variations in axis shifts. Based on the investigation, the optimal aerodynamic properties were found to be a rotor spacing of 3.6, which showed increased thrust and reduced power consumption. The study also showed that the aerodynamic performance of the quadrotor might be enhanced by mild aerodynamic interference, especially at a 3.6 rotor spacing. In the end, the results indicate that taking into account the distance between rotors and the resulting aerodynamic interference is the best rotor configuration for increased aerodynamic performance, demonstrating the possibility of improving the efficiency and design of quadrotor systems.

The influence of interaction between rotors on the aerodynamic performance and aeroacoustics characteristics of mini-UAVs was examined by Zhou, W. et al., (2017). The results showed that although the distance between rotors had no effect on the rotor's thrust coefficients, it did cause a considerable rise in thrust fluctuations. Thrust fluctuations were found to be 250% more in the twin-rotor scenario than in the single-rotor scenario. This was linked to intricate flow interactions that were identified by PIV and Stereoscopic PIV observations. Furthermore, increasing the aeroacoustics noise levels and intensifying force fluctuations were seen upon decreasing the separation distance. The L = 0.05D example showed a maximum increase in aeroacoustics noise of about 3 dB when compared to L = 1.0D. This increase was ascribed to strong thrust variations and turbulent flow interactions inside the rotors. The study highlights the significance of taking these interactions into account in the design and operation of such systems by offering insightful information on the performances of tiny UAVs affected by rotor-rotor interactions.

Herz, S et al., (2023) investigated the impact of interactions between rotor-rotor and rotor-body on multirotor vehicles performance. Previous understanding of these interactions was least to numerical analysis and testing in controlled environment, leaving the effects on full-scale multirotor flight vehicles largely unknown. The study utilized different flight test and wind tunnel experiments of a quadrotor at various speeds and the angles of sideslips to calculate these interactions. The results indicated a decrease in aft rotor performance when in the cross configuration with the fuselage, as well as a decrease in thrust coefficient with the fuselage in the plus configuration. Additionally, inwardly rotating rotors resulted in increased thrust coefficient compared to outwardly rotating rotors. The study found that rotor performance trends between the two rotor configurations were nonlinear with respect to the velocity of the vehicle and sideslip angle. and additional tests at intermediate angles confirmed these trends. The data also showed an increase in rotor thrust coefficient with vehicle forward flight speed up to 12 m/s but performance degradation for the aft rotors beyond that speed. The findings highlight the significant impact of rotor-rotor and rotor-body interactions on the performance of multirotor vehicles, providing valuable insights for the design and operation of such vehicles.

In their evaluation of the impact of interactional effects on quadrotor efficiency during forward flight, Kostek, A.A. et al., (2024) took into account a variety of hub spacings, including overlapping blades, forward and backward tilt angles, and square and diamond designs. The analysis is based on simulations and data from wind tunnels using four high-fidelity tools and four mid-fidelity computational methodologies. The results indicate that for non-overlapping rotor spacings, the efficiency of a diamond configuration is improved by 5% when compared to isolated rotors. Conversely, interactions in square alignments are negative for all test scenarios that were analyzed. The ideal arrangement has a blade overlap of 0.04D. Backward rotor tilt exhibits a more marked tendency with stronger interactions. In consideration of this, the diamond configuration's efficiency rises by 11% at 1.2D rotor spacing. For forward rotor plane tilt, computational findings correspond well with measurement data. However, for backward tilt angles, there is a scattering in computed values, particularly for torque, but overall trends are preserved. According to the study, rotor phasing has very little effect on quadrotor efficiency.

Using numerical simulations, Dougherty, S. et al., (2024) investigated the effects of propeller spacing and onset flow conditions on the effectiveness of quadcopter propellers. A quadcopter arrangement with high spillover propellers and a two propellers setup with propellers distanced perpendicular to the beginning flow were the two configurations studied. Propeller performance was identified to be barely affected by the axis shift, according to the simulations of twin propeller. Based on a comparison with previous research, propeller dispersion was found to have a considerable impact on the amount of wake interaction between neighbouring propellers in still air, which in turn affected propeller performance. According to the quadcopter simulations, when an onset flow was applied parallel to the propeller spacing, propeller pairs' performance changed noticeably, and the velocity of this onset flow affected propeller performance more than the flow's turbulence level. Notably, the study provides significant insights for the dynamics of multirotor flows and contributes to enhancing the understanding of quadcopter propeller performance under different flow conditions and separation distances.

2.3 STUDIES ON THRUST BY ROTORS IN UAV

Examining the thrust generation of propeller blade width-spans with and without mouth-ring application was the goal of Othman, M.N.K. et al., (2019). Testing was done with and without a mouth ring in two different situations. The thrust was computed using velocity, voltage, and rpm. In the tests, two propeller sizes were used, and the mouth-ring application was determined by measuring the gaps 10, 8 and 5 millimetres between the mouth-ring wall and the blade tip. According to the results, the 10mm mouth-ring generated more force than the other gap sizes. Furthermore, the application of the mouth-ring showed superior thrust creation as compared to its absence. Based on the results obtained from the experiments, the study finds that the mouth-ring application is useful for improving thrust generation.

Dahal, C. et al., (2021) looked into how lower air density affects thrust generation, which limits the use of available unmanned aerial vehicles (UAVs) for the missions of search and rescue at high altitudes. The goal of the project is to use lightweight aluminium alloy 1,060 to design and analyse a thrust-optimized blade for altitudes between 3,000 and 5,000 m. To evaluate the blade's performance within the altitude range, a design and analysis algorithm based on blade element theory was created. At 70% of the blade span, the proposed blade discovered a total stress of 6.0 MPa, which is well within the aluminium alloy's yield strength. According to modal analysis, the blade's safe operating range is 0-5,000 RPM. A force of 0.92 N at 2,697 RPM was obtained via experimental analysis at 1,400 m altitude, while an analytical solution predicted 1.7 N with 85.6% efficiency. The experimental results were confirmed by computational fluid dynamics (CFD) analysis in ANSYS CFX, which displayed a thrust value of 2.27 N under the same circumstances. The study comes to the conclusion that the planned blade is structurally safe for operation at high altitudes, potentially saving lives during search and rescue operations, especially in the Himalayas.

The effectiveness of a 10*4.7-inch curved tip propeller was studied by Chakchouk, M.A. et al., (2023) as a potential replacement for ducted propellers in circumstances when their use is restricted by size or mass. The study models the propeller's performance using Computational Fluid Dynamics and compares the outcomes of the simulation with experimental data obtained from the University of Illinois at Urbana-Champaign. One important performance metric that is employed is the thrust coefficient. Both advantages and disadvantages of the curved tip propeller are revealed by the study, especially when considering the mission profile and battery life of a quadcopter unmanned aerial vehicle. The study addresses the advantages and disadvantages of a curved tip propeller in the mission profile and is referred to as the first research of this type on a drone. According to the computational research, curved tip propellers perform better across the whole advance ratio (j) value range, potentially resulting in a 25% increase in thrust force. The propeller is more competent with the same size because of the increase in thrust force that enables a bigger weight to be lifted when hovering. Nevertheless, the drone's flight time is reduced by 30% as a result of this increased thrust.

Overall, the study recognizes the trade-off of shorter flight times while highlighting the possible performance benefits of curved tip propellers.

Ahmad, F et al., (2021) found the propeller thrust coefficient (Kt) through flow analysis of a quadcopter propeller is the primary goal of the current work. Given that the quadcopter propeller opposes both aerodynamic and gravitational forces as a result of drag and weight on it, flow analysis is crucial. Thus, for fluid flow simulation, a 3D CAD model of a quadcopter propeller was created in Catia and loaded into Ansys fluid flow (fluent) module. Using a k-epsilon turbulent model, the flow over a propeller was simulated at various angular speeds. The acquired findings indicate that the propeller can produce the necessary thrust force to lift larger cargoes, according to the design. Additionally, the thrust coefficient was determined by utilizing the generated thrust force at various angular speeds from 4000 to 11500. if the quadcopter is made of CFRP, then the thrust produced at the speed of 4000 was not enough since it was only 8.44 N and the required thrust is 9.29N. However, from 5000 rpm, the thrust produced is increased and enough to lift the weight.

In their study from 2022, Kišev, M. et al. talked about how crucial it is to monitor things like thrust, current consumption, and the maximum rotational speed of the propellers used in unmanned aerial vehicles (UAVs). The accuracy and data storage of current measurement techniques are frequently constrained. The goal of the article is to create an all-purpose experimental measuring system that can measure temperature, humidity, and battery voltage in addition to other parameters for more accurate readings. The system securely collects, measures, and a micro-SD card is used to store the data using a processing circuit of a ATmega2560 microcontroller. Using this technique, the scientists were able to monitor the propeller's dependency on different input conditions and analyse its properties. The measurements showed that cubic functions may be used to approximate the dependencies, which makes it possible to predict how the propeller will behave under unmeasurable circumstances. The article introduces a custom measurement board that can collect, process, and store data for later use, as well as the construction of an experimental platform for measuring the thrust of UAV propellers. The system was put to the test using a Dragonfly 1755 propeller with unknown characteristics and a Tarot 5008 340 KV BLDC motor, yielding important insights into their real-world behaviour.

2.4 STUDIES ON MATHEMATICAL MODELLING IN UAV

Using MATLAB/Simulink software, Jenie et al. (2020) provide a thorough mathematical modelling and simulation of a quadrotor unmanned aerial vehicle (QUAV). The major goal of the research is to offer a trustworthy mathematical model for designing and assessing the control systems needed to keep a QUAV stable. The modelling results shed important light on the QUAV's dynamic properties. The authors noted the logical commands needed to change the rotors' thrust, the rotors' limitations in processing multiple commands at once, the automatic control systems' reliance on the behaviour of the vehicle, and the interdependence of the various degrees of motion. In the simulations, the constructed mathematical model and the related control systems performed satisfactorily, suggesting the model's potential as a basis for more QUAV analysis and control design. The study highlights how crucial this research is to improving our knowledge of and ability to operate QUAV systems, which are finding more and more uses in civilian applications.

Benić, Z. et al., (2016) investigated the creation and use of a mathematical model for a quadcopter, which has four propellers. The study's main goal is to create a mathematical model that precisely captures the quadcopter's behaviour and can be adjusted for other multirotor aerial vehicle combinations. The importance of this mathematical model for modelling the control algorithm and quadcopter movement is emphasized throughout the paper. The research then explores the dynamics of quadcopters, applying the Newton-Euler method to incorporate forces and torques into the model. The final derived equation, which unifies the quadcopter's mathematical behaviour, represents Newton's second law in matrix notation. The authors establish a hybrid coordinate system and construct dynamic equations with regard to it in order to simplify the model. Then, simulations are run using these equations. Based on four different scenarios, the study shows graphical depictions of the quadcopter's behaviour. In each scenario, the revolutions per minute (RPM) of the propellers are adjusted to provide controllable variables that enable one of the four basic quadcopter movements in space. The graphical representations provide information about how the quadcopter responds to various RPM settings and show how this affects its space movements.

The design and study of a miniature indoor multirotor unmanned aerial vehicle (UAV) was the main topic of Kotarski, D. et al., (2017). Unmanned aerial vehicles with multirotor (UAVs) are frequently utilized for a variety of purposes, including monitoring, assessment, and disaster site observations. The study's micro indoor multirotor is made up of a frame, a propulsion system, sensors for urban flight, and an open-source autopilot that can be used to create control algorithms. A thorough nonlinear mathematical model that falls under the category of rigid body dynamics is presented in this study. To sum up, this study provides an in-depth examination of the design factors and analysis of a small indoor multirotor unmanned aerial vehicle. It offers insightful information on how FMRC configurations are developed and what makes them perform. In order to make well-informed decisions when choosing and optimizing UAV designs for indoor applications, it is essential to analyse and comprehend the many attributes and capabilities of these multirotor configurations. This may be done with the help of the derived CAS matrix for

FMRC. The comprehensive simulations, practical observations, and mathematical modelling advance our knowledge of multirotor UAV performance and lay the groundwork for future studies in this area.

The creation of a mathematical model for a quadrotor UAV in civil is discussed by Jenie, Y.I. et al., (2018). Despite the widespread use of quadrotors, the authors point out that appropriate techniques for their mathematical modelling, design, and analysis are still lacking. In order to close this gap, the study describes a mathematical modelling procedure for a quadrotor unmanned aerial vehicle (UAV) employing a methodology identical to that of modelling traditional aircraft. A state space format that accurately captures the vehicle's dynamic characteristics while providing a linear description of the forces and moments operating on it is the aim of this technique. Additionally, using standard techniques in flight dynamics and control, the research examines the mathematical model's controllability and stability. Understanding the behaviour of the quadrotor under various flight conditions and how to control it successfully depend on this analysis. The authors hope to shed light on the Quadrotor's overall performance and manoeuvrability, as well as how it responds to different inputs and disturbances, by looking at the model's stability and controllability.

Six state variables that affect the quadrotor's stability, or equilibrium point, were chosen by Vanitha, N.S. et al. as the state vector and control vector. These state variables include things such as the quadrotor's position, orientation, and velocities. The MATLAB Simulink environment was used to simulate the state space model, and the outcomes were examined and verified for each of the six state variables. The paper's main conclusion is that the quadrotor's stability point is significantly impacted by changes in these six state variables. By employing state space methodologies to precisely describe the quadrotor, the researchers were able to improve their perception of the variables affecting the control and stability of these compact yet adaptable aircraft.

2.5 STUDIES ON PAYLOAD TRANSFER IN-BETWEEN UAVS

The design and implementation of a system for multirotor aerial vehicles to exchange payloads while in the air was investigated by Shankar, A. et al., (2020). The suggested method tackles issues including closeness, contact forces, and environmental uncertainties while making use of multirotor vehicles' hover capabilities. The technology doesn't need special flight manoeuvres and is made to be portable to a variety of vehicles. In-depth test results showing two vehicles sharing a tiny, 60g payload unit between them are presented. The suggested method is widely applicable to a range of multirotor since it does not call for specialized flight controllers and entails the controlled swinging of a payload system on both the sender and recipient vehicles. It tackles problems with vehicle separation, relative positioning accuracy, and contact forces. The payload and disc are kept in phase throughout rotation by means of a visual feedback system and a dependable quick-release connector in the system. The outcomes of the experiment show that the system can cause and maintain the payload's rotation in spite of external disruptions. The system's ability to adjust to various climatic conditions and characteristics is demonstrated, along with its capacity to transfer payloads between vehicles while in hover as shown in Figure 2.1. A real-time dynamic programming (RTDP) method was presented by Mohiuddin, A. et al., (2022) to optimize the joint delivery of a payload with two unmanned aerial vehicles. Simplifying the system model results in a cost function that can be decreased to cut down on travel time or energy use. Depending on a control factor, the algorithm can alternate between different optimization techniques. According to the authors, this is the first time that load distribution and lateral travel optimization have been used in tandem with UAV payload delivery. In the cost function, a term is added to account for the lift distribution among the UAVs.

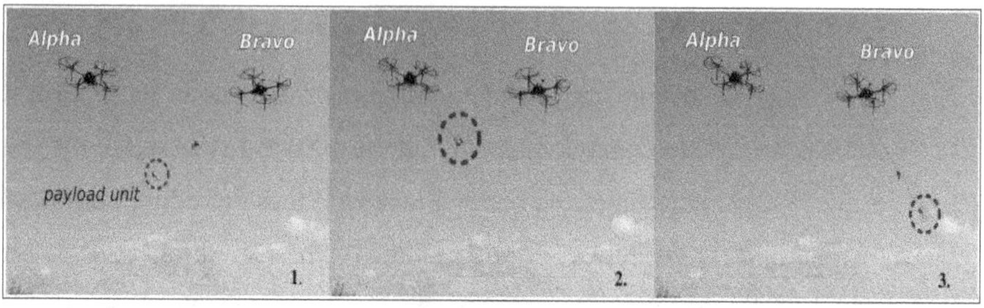

Figure. 2.1. Payload Transfer Demonstration (Hanging Model)

According to the study, raising the control factor shortens the travel time but decreases energy consumption; this might be adjusted depending on the purpose of the UAV. The discretization path straight line without taking complicated paths or barriers into account has limitations. Future studies could involve creating paths free of obstacles, optimizing the velocity along the path, and testing forward flights at high speeds experimentally.

The attempt was made to apply inverse reinforcement learning (IRL) for drone swarm analysis, which was described in this article by Kim, K.J. et al., (2019). They found that more diverse scenarios produced larger entropies, better performance, and more interference through the application of reward distributions found by IRL to compute and analyze entropy and interference. Furthermore, different patterns are noticed in entropy and interference when one of the two variables command style was heterogeneous and the other was homogenous. Correlation analysis was used to investigate the dual effect of interference in various drone scenarios and found that both the command cognitive style and war strategy contributed both favourably and unfavourably. Two quadcopters were utilized by Prajapati, P. et al., (2020) to investigate the issue of jointly conveying a cargo suspended by a cable. Unlike previous quadcopter transportation research, which concentrated on designing of complex routes and autonomous control, a master-slave strategy is used in this paper, and the slave quadcopter stabilizes the payload oscillations, and a human controls the master quadcopter. When a cable-suspended payload system of two quadcopters is connected and under actuated, manual control becomes difficult. On a manifold, Lagrangian mechanics is used to obtain the motion equations and linearization of the system is done. They created a controller based on Lyapunov that will simplify human control of the master quadcopter and lessen payload oscillations during transit.

Figure. 2.2 Payload Exchange Demonstration (Rigid Type)

Unmanned aerial vehicles with several motors were used by Shankar, A. et al. to complete a variety of tasks. However, the cargo can only remain in the air for so long due to the UAS's flight endurance. The authors of this work suggest a novel way to move a payload between two multirotor in flight while keeping it in the air and stationery. Based on a visual-feedback and grasping pipeline, this technology enables UAS to act as interchangeable carriers by enabling one UAV to capture the cargo being carried by another. By attaching the payload outside along a single, rigid link and allowing the UASs to roam around it, they were able to keep it operational during the transfer to a new carrier. Through this work, a novel method that allows two multirotor to exchange cargoes while still operating simultaneously has been introduced. By using the chosen method, one multirotor merely needs to stay still while the other extends a stiff link for it to grab as a support bar. A transfer can be performed with exact visual servicing and active grabbing without disturbing the payload which is shown in Figure 2.2.

2.6 PROBLEM STATEMENT

Drone delivery provides a quicker, more environmentally friendly, safer method of obtaining essentials in any place. However, the usage of drones has been increasing now a days, due to the above discussed challenges in chapter 1, it has not been reached to be the common tool for longer area coverage. Among the all-other challenges, the interference effects playing a dominant role in the overall performance including safety factors. During the swam condition also, it majorly impacts the performance, robustness, resource utilization, scalability, safety, and adaptability across various applications and environments. If the closer proximity between the drones without diminishing the thrust is found, it would be more beneficial and feasible to carry out the multi-drone operation for longer coverage.

Other than the conventional single drone operation, if we employ multiple drones for any specific purpose, it would be more advantageous such as increased speed and accessibility, efficient time management, access to remote Areas, reduced traffic congestion and lower costs. Despite the fact that the research in multi-drone operation is becoming popular and needful, the closer proximity between drones is not discussed in any other earlier research study in both the axial and lateral direction. This research concentrates on conducting experimental and numerical study for understanding the interference effects and finding the minimal operating distance between two rotors.

CHAPTER - 3

METHODOLOGY

The outline of the iterative and interconnected nature of engineering research, emphasizing the integration of theoretical models with practical experiments to explore complex systems such as drone dynamics is shown as a flow chart in Figure 3.1

The first step in the research involves defining the problem to be addressed, guided by a literature review to understand existing knowledge and gaps. The project includes three parallel tracks such as 1. Experimental Evaluation 2. Numerical Evaluation and 3. Mathematical Modelling.

Experimental Evaluation involves developing a test rig, selecting appropriate propellers, and deciding on their spacing (axial or lateral). This setup helps to study the interaction effects of propellers. In Numerical Evaluation, a 3D scan of the propellers chosen in the experimental setup is used to capture the geometry, which helps in configuring the simulation settings. Mesh generation for the computational domain and setting up turbulence modelling and boundary conditions are critical for accurate simulation results. Once the convergence is reached in simulation processes, it is aimed to visualize flow patterns and other performance metrics.

Mathematical Modelling involves building models for drone dynamics and flight control, which are crucial for simulating drone behaviour under different conditions. The simulated results accuracy is then confirmed by validating them with experimental results. This is the crucial step to confirm that the mathematical models and simulations faithfully represent real-world dynamics. Finally, the research findings are discussed, results are analysed, and conclusions are drawn. Recommendations for future work or applications in drone technology are also typically provided.

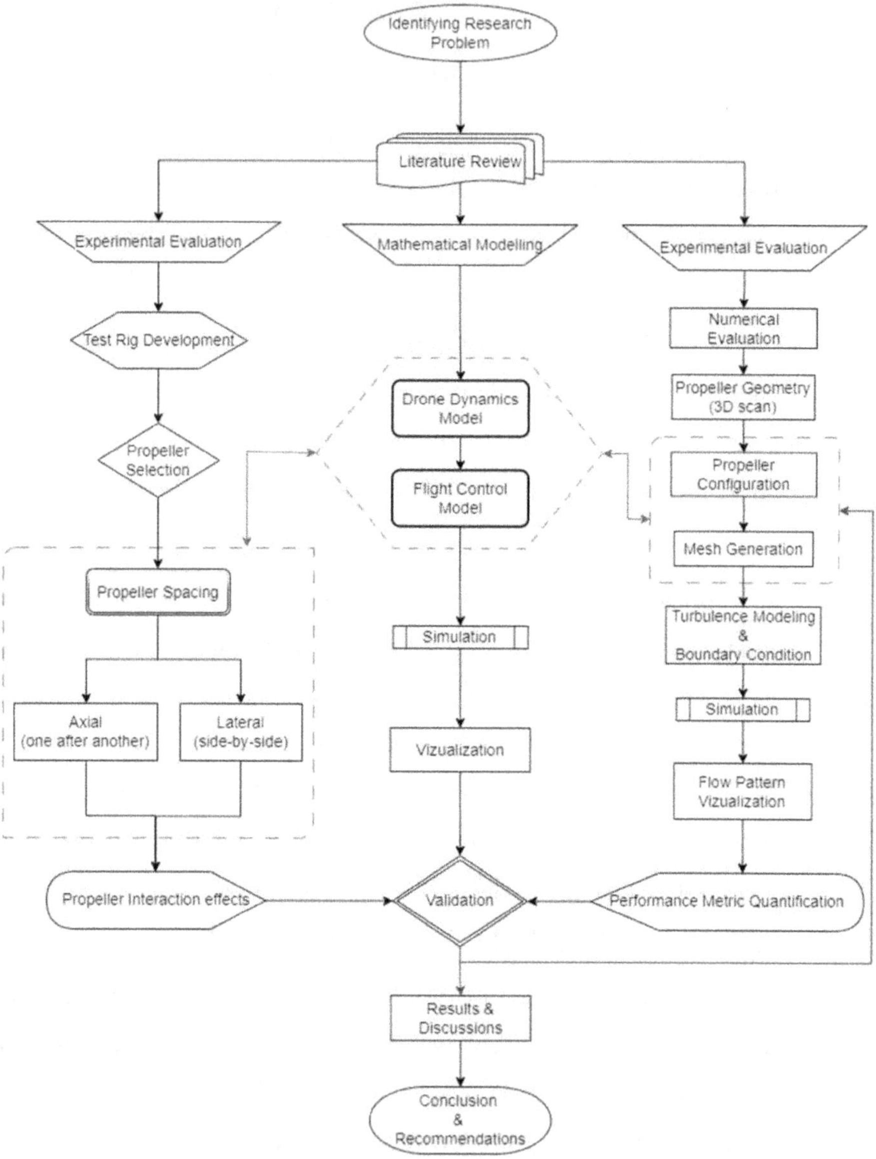

Figure 3.1 Methodology Flowchart

3.1 EXPERIMENTAL EVALUATION

3.1.1 Design and Development of Propeller Interference Test Rig

The propeller is attached to the motor shaft which is mounted with a Thrust stand (TS1520), The mechanical framework of the test rig allows the operator to move and regulate the position of the Thrust stand. The RC benchmark GUI on the PCs can be used to gather motor RPM and thrust statistics. These characteristics enable the study of UAV rotor-to-rotor interference effects. The mechanical frame was made up of stainless steel with the dimensions of 157cm, 60cm and 70cm as the length, breadth and height of the base support member. There are two thrust stands each one is mounted on a vertical slider (S7, S8) that can move vertically so that the height can be set, the slider S7 and S8 is mounted on the vertical member which is welded on the Axial slider (parallel to the propeller axis) S5 and S6 respectively. The axial slider runs through a lateral member attached to the lateral slider pair S1, S2 and S3, S4 respectively on both sides to make the base rigid. Two lithium polymer four cell battery (14.8V, 2200mAh) are used to power two propellers and the used Propeller, Battery, and Thrust Stand are shown in Figure 3. The Thrust Stand Specifications are listed in Table 3.1.

Table 3.1 Thrust Stand Specifications

Name	Limit
Thrust	-5kgf to 5kgf
Voltage	0 V to 35 V
Current	0 A to 40 A
Angular Speed	0 eRPM to 190000 eRPM

3.1.2 Selection of Propeller

The 13-inch (EOLO 1350) propeller was selected for conducting experimental and numerical tests. There are many positive reasons for having taken this propeller. The advantages are as follows

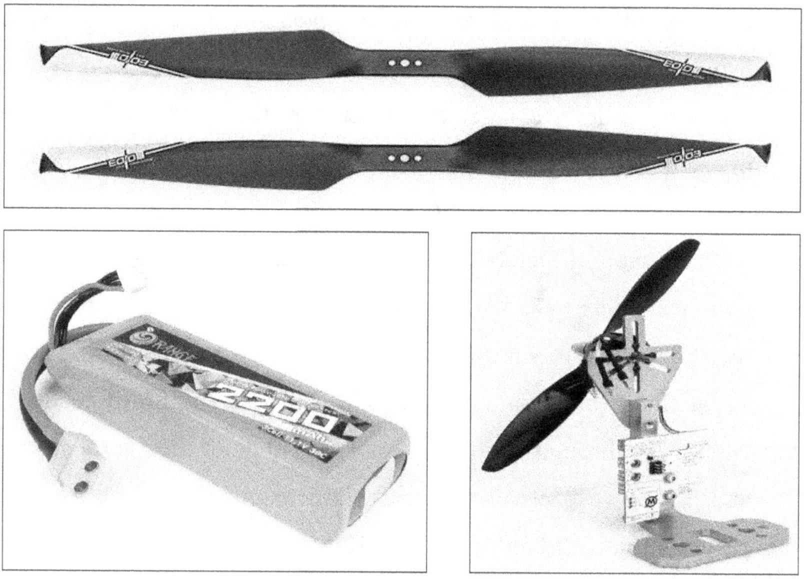

Figure 3.2 Components of Test Rig-Propeller(top), Battery (left), Thrust Stand (right)

Efficiency: Designed specifically for quadcopters and multirotor, these propellers offer superior lift and stability in the air. Their construction guarantees the best possible thrust-to-power ratio, improving battery life and enabling your drone's motors to run smoothly.

Material: The carbon fibre-reinforced nylon used to make the EOLO propellers creates a compromise between weight and strength. Their performance and durability are enhanced by this combination of materials.

Compatibility: EOLO propellers can be used with a variety of UAVs, making them adaptable to various models and setups.

Dimensions: Measuring 13 by 5 inches, their size makes them ideal for a range of UAV uses. The proportions offer a nice balance between mobility and lift capacity.

3.1.3 Experimental Test

The experimental tests were carried out by using the interference test rig shown in Figure 3.3. The thrust produced by individual rotors at various speeds from 4000 rpm to 8000 rpm with 1000 rpm intervals and at different spacings can be measured using experimental models.

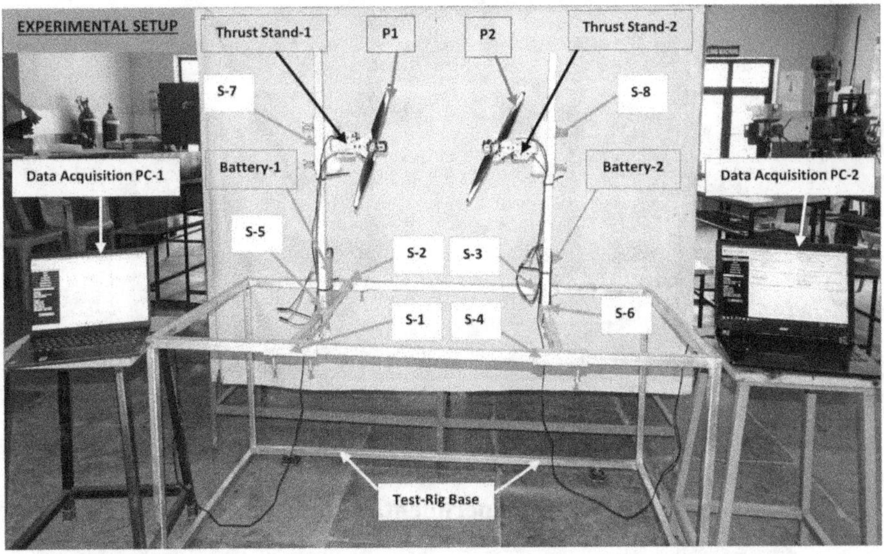

Figure 3.3 Photographic view of Experimental Setup

The measured thrust of each rotor is used as a quantifying parameter of interference effects. Interference between the rotor vortices is indicated by decreased thrust at closer spacing.

3.1.4 Lateral Spacing in Experimental Setup

Two similar rotors are positioned side by side and the rotating direction is opposite to each other which makes up the lateral test configuration shown in Figure 3.4. To measure thrust, each module of the rotor assembly is equipped with a 5-kgf range load cell. The spacing between rotors kept in tests is termed as Lateral swift (LS). In axis swift LS2.2, the distance between the rotors is kept at 2.2 times the radius of the rotor from the rotational axis of them. Similarly, LS2.4, LS2.6, LS2.8, AS3, and LS3.2 distances were also used.

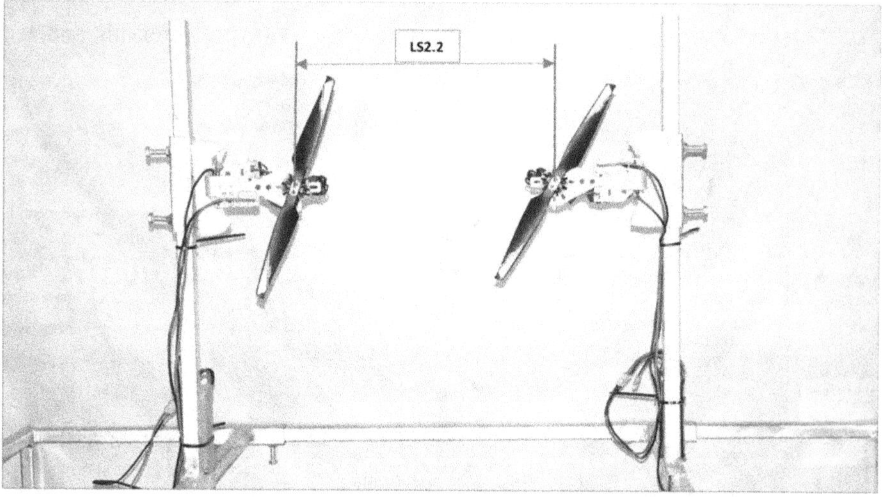

Figure 3.4 Photographic view of Lateral Spacing (LS2.2)

3.1.5 Axial Spacing in Experimental Setup

Two identical rotors positioned one after another and rotating in opposite directions make up the lateral test configuration shown in Figure 3.5. To measure thrust, each module of the rotor assembly is equipped with a 5-kgf range load cell. The spacing between rotors kept in tests is termed as axis swift (AS). In axis swift H1.5-V1.5, the distance between the rotors is kept at 1.5

times the radius of the rotor from the rotational axis of them. Similarly, H1-V1 and H0.5-V0.5 cases were also tested. In each AS, the distance between one rotor disc to another rotor disc in axial direction was 0.5R, 1R and 1.5R.

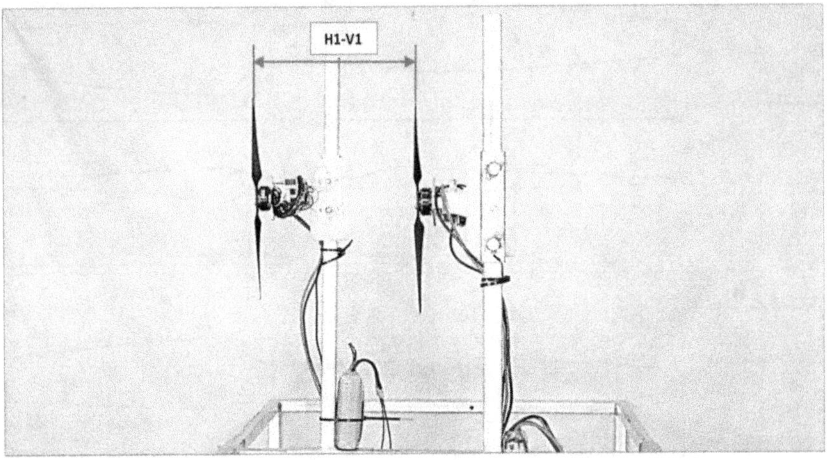

Figure 3.5 Photographic view of Axial Spacing (H1-V1)

3.2 NUMERICAL MODELING

3.2.1 Design and Development of Numerical Model

Analysing aerodynamic performance through flight tests and visualizing wakes is difficult due to the complexity of the aerodynamic interference between the rotors. Therefore, the most efficient way to visualize for a broad range of configurations is the CFD simulation. Rotor vortex interactions are taken into account in numerical models. The complex flow fields produced by rotors in interaction with one another are simulated by these models. Understanding interference effects requires analysing wake patterns between each rotor. Interactions between wakes affect stability and efficiency. The test model that was used for the experimental tests was scanned to get the 3D model that can be used for the numerical analysis. The dimensions of the

EOLO propellers 13x5.5 is shown in Figure 3.6 and the Figure 3.7 shows the scanned 3D model.

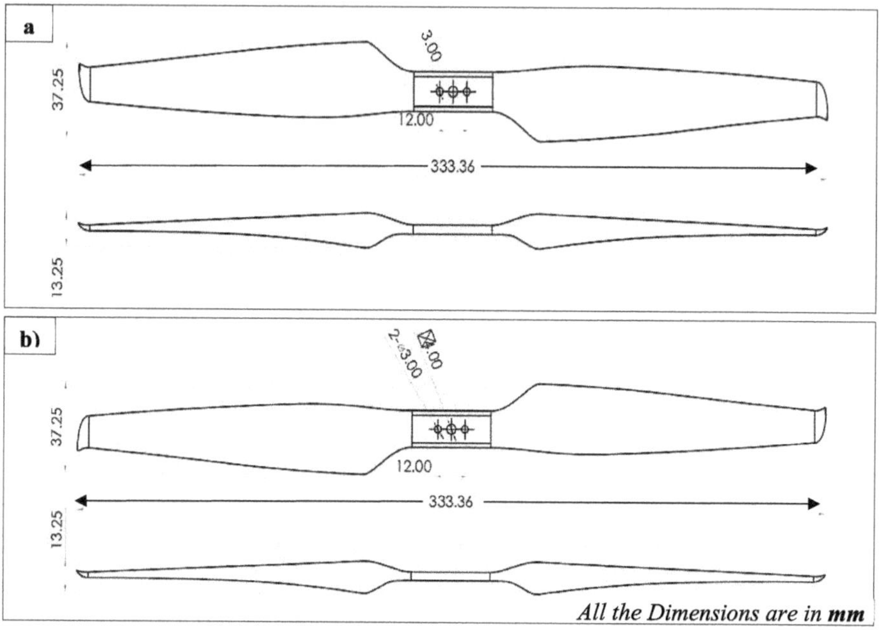

Figure 3.6 Orthographic View of Propeller a) counter-clockwise propeller, b) clockwise propeller. *Resource: https://sunnyskyusa.com/products/eolo-cn13-5-prop*

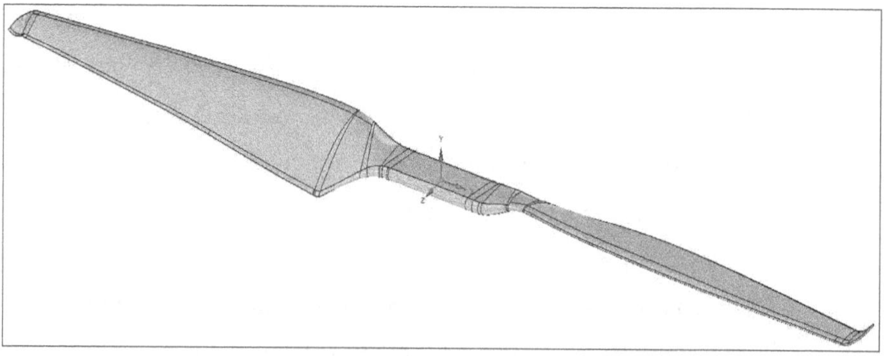

Figure 3.7 3D Scanned Propeller (EOLO Propeller-13x5.5)

3.2.2 Boundary Conditions

The numerical analysis was performed at various RPMs, including 4000, 5000, 6000, 7000, and 8000 rpm. The remaining constant in the rotating coordinate system was the propeller's flow field. The governing equation was solved using the pressure-based solver after the flow field had been discretized using the finite volume method. Figure 3.8's grid is completely tetrahedral in both the static and revolving regions. Tetrahedral components are easy to compute, which reduces processing costs, and can be a great fit for geometries of any shape. Fifteen levels of inflation were modelled to capture the boundary layer effects. SST k-ε was the chosen as turbulence model for the analysis. At the chosen inlet, the velocity was just 1 m/s. 25 °C temperature and 1.22 kg/m^3 density were used for each simulation. VA = 1 m/s is the preset advance speed. A static analysis was performed based on pressure. The prerequisites for defined boundaries are as follows. The turbulence intensity was set as 1% for inlet boundary and velocity inlet. On the boundary of outlet 0 Pa was defined as reference pressure with the static outlet pressure. For every solid surface, the no-slip condition was given and surrounded the propeller domain, free slip wall condition is given. The front and back face of the streamwise rotating domain are the sliding interfaces and these are used for transferring the flow features between the stationary and rotating domains. The same numerical test cases were carried out as followed in the experimental tests in both the axial and lateral spacing conditions. The numerical model for the respective cases is shown in Figure 3.9

3.2.3 Lateral Spacing in Numerical Model

In lateral spacing, two identical propellers are laterally placed in the same rotating plane and the left and right rotor rotate in clockwise and counter-clockwise direction respectively. The distance between the axis of the propellers called Axis Shift or Lateral Spacing are varied. It varies from LS 2.2 to LS 3.4 with 0.2 interval. For each spacing, numerical tests are conducted from 4000 RPM to 8000 RPM in 1000 rpm interval. Figure 3.10 shows the meshed model of Stationary and Rotating domain of LS-2.2 Configuration

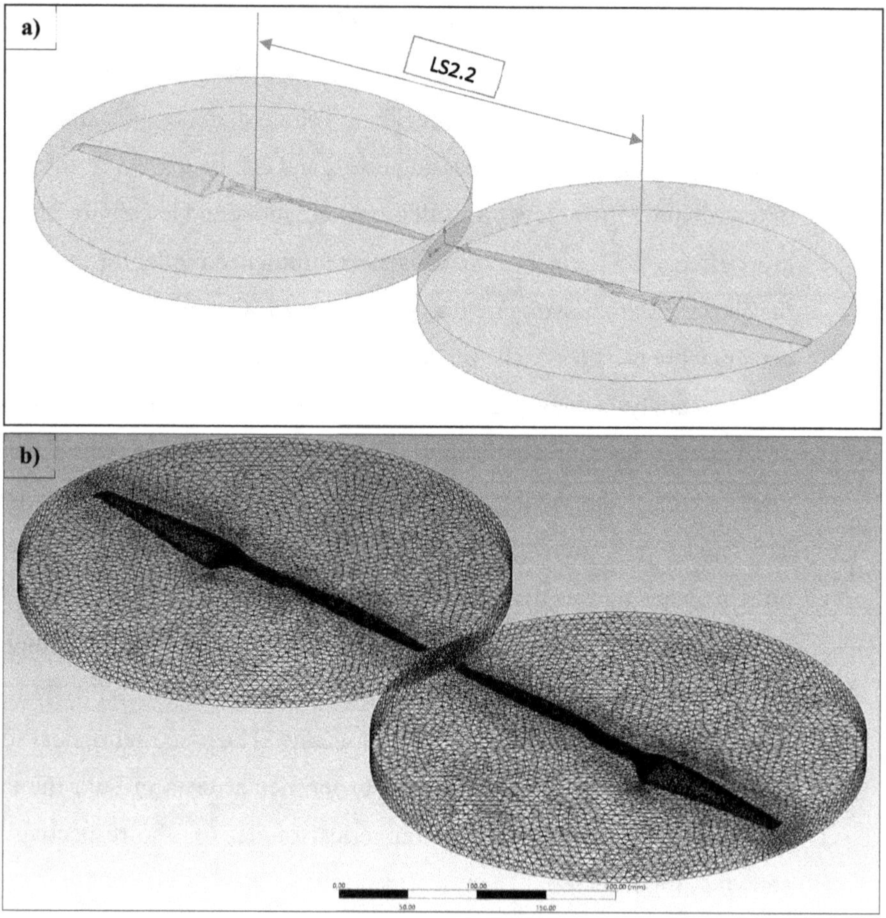

Figure 3.8 Lateral Spacing LS-2,2 Configuration-Rotating Domain a) Geometry of Rotating Domain b) Rotating Domain Mesh

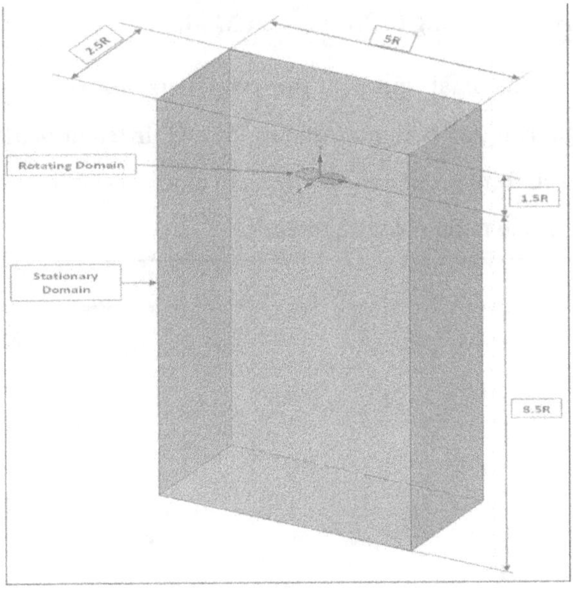

Figure 3.9 LS-2.2 Configuration-Stationary and Rotating Domain

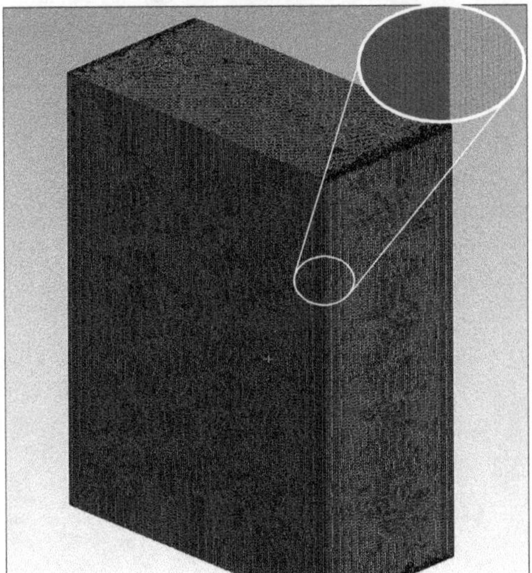

Figure 3.10 LS-2.2 Configuration-Stationary and Rotating domain Meshed model

3.2.4 Axial spacing in Numerical Model

In axial spacing, the propellers are placed in one after another arrangement and both are rotating in two different planes as shown in Figure 3.11. The front propeller P2 rotates in clockwise direction and the aft propeller P1 rotates in anticlockwise direction.

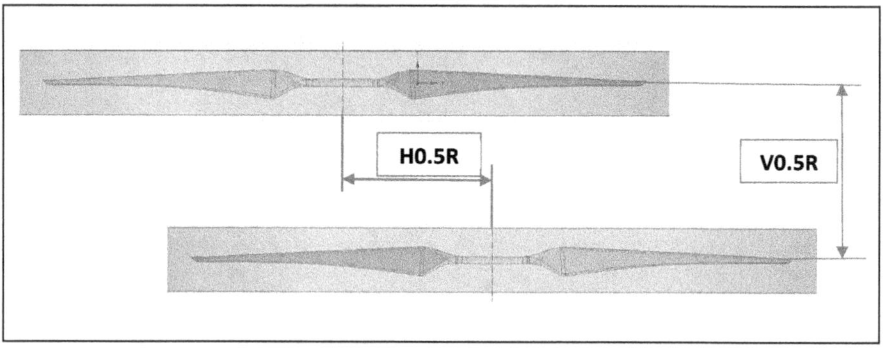

Figure 3.11 Axial Spacing H0.5R-V0.5R configuration-Rotating Domain

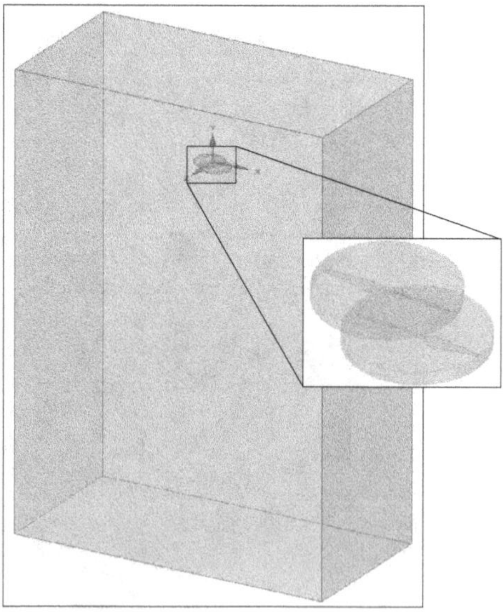

Figure 3.12 H0.5R-V0.5R Configuration-Stationary and Rotating Domain

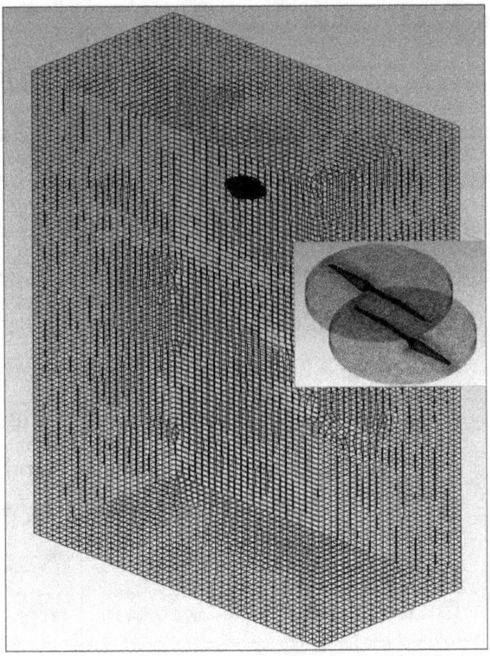

Figure 3.13 H0.5R-V0.5R Configuration-Stationary and Rotating Meshed model

For conducting the numerical tests, two propellers are moved in both the axial and lateral direction with certain distance. when the propellers are in H0.5-V0.5, it means that lateral axis distance between the propellers are 0.5R denoted as H0.5 and the distance between each rotating propeller plane in axial direction is 0.5R mentioned as V0.5. Similarly, simulation is carried out for H1-V0.5, H1.5-V0.5, H0.5-V1, H1-V1, H1.5-V1, H0.5-V1.5, H1-V1.5, and H1.5-V1.5. Figure 3.12 shows the geometry of the rotating domain kept inside the stationary domain and Figure 3.13 the meshed model of Stationary and Rotating domain in H0.5-V0.5 Configuration.

3.2.5 Mesh Convergence Study

At a speed of 5000 rpm, the two axial and lateral test cases were used for the grid independence investigation. The AS2.2 test case was used for the lateral direction, the AS0.5 test case for the axial direction, and there was 0.5 spacing between the rotor discs. Table 3.2 lists the specifications of the simulation that was conducted using the intermediate and medium-level mesh elements, the results verifies that the difference is very minimum.

Table 3.2 Mesh Convergence study results

Propeller Configuration	Mesh	Stationary domain		Rotating domain		C_T
		Number of Nodes	Number of Elements	Number of Nodes	Number of Elements	
Lateral Spacing AS2.2	Coarse	4965726	26384627	1012589	6245837	0.02103
	Medium	5100231	28847836	1177642	6505436	0.02126
	Fine	5354685	31597452	1386499	6815862	0.02128
Vertical Spacing AS 0.5	Coarse	879141	4522181	0983256	6336518	0.02099
	Medium	880751	4831728	1183578	6538153	0.02100
	Fine	892611	5042533	1484179	6740751	0.02121

The produced thrust coefficients are matched with the experimental results and the difference between the course, medium, and fine mesh elements are least at the level of 0.0001. As such, for further numerical modelling, the medium size mesh elements are taken.

3.2.6 Turbulence Model Study for Validation

The thrust coefficients for various turbulence models were compared with the thrust coefficient obtained by the experimental results using the following formulas to determine the percentage of variance in CT (ΔCT).

$$\Delta C_T(\%) = \frac{C_{TCFD} - C_{TEXP}}{C_{TEXP}} * 100$$

Table 3.3 Turbulence Model

S.No	Model	ΔC_T (%)
1	Standard k −ε	3.3901
2	SST k −ω	2.50723
3	Realizable k −ε	2.67648

The obtained ΔCT values for the tested turbulence models of Two-equation SST k −ω, Standard k −ε, and Realizable k −ε were 2.5, 3.39, and 2.676, respectively which are listed in Table 3.3. The SST k −ω method with the minimum level variation is taken for further tests and analysis.

3.3 Development of Mathematical Modelling

The physics and mathematics model of a quad-propeller drone was developed for the Tarot FY650 carbon frame. Based on the mathematical model, a SIMULINK model was developed to mimic all aspects of the drone's dynamics, including its location and the incorporation of gusts to confirm its stability. A numerical simulation of the system's control response was performed, and the results were analysed. A mathematical model for a distinct kind of quadcopter model, each with its own set of parameters, may be rapidly constructed using the phases.

3.3.1 Mathematical Modelling of TAROT 650

A mathematical model of the quadcopter must be derived before the hardware's simulation data can be quickly analysed. Figure 3.14 shows the 3D model of TOROT 650 quadcopter frame whose parameters are used to develop the mathematical model. The research done by M. A. Alsharif et al. (2017) had a considerable influence on the development of the quadcopter UAV model used in this paper. A four-arm and a symmetrical structure quadcopter is considered. Every motor was assigned to the distant extremity of the arm, and the Electronic Speed Controller, also known as the ESC, was placed beneath the middle of the arm.

Figure 3.14 3D CAD Model of Tarot 650

3.3.2 UAV Vehicle Dynamics

Studying the effect of the rotation of individual propeller disturbs the movement of a quadrotor is the most logical way to understand its motion. When all rotors spin at the same speed, equal forces are produced. By adjusting the rotor speed, the quadrotor can move along the z-axis and achieve a hovering position or state at a specific propeller speed. Movement along the x_b and y_b axes occurs when two adjacent rotors rotate faster than others rotors. A quadrotor's rotors apply torque to the body in the opposite direction of their rotation. The torques of two rotors cancel each other out when they rotate at the same rpm in opposite directions. By adjusting the speed of the diagonal rotors, the yaw angle (represented by the symbol ξ_E) can be controlled, causing the Z_b

axis to move in a positive yaw direction. The symbol E is used to represent the earth's initial frame, and the body-fixed frame by the B symbol. The linear position vector of the x, y, and z axes of the earth's inertial frame is defined in equation 3.1.

$$\xi_E = [\, x \;\; y \;\; z \,]^T \tag{3.1}$$

$$v_E = [\varnothing \;\; \theta \;\; \Psi \,]^T \tag{3.2}$$

$$V_b = [\, u \;\; v\, w \,]^T \tag{3.3}$$

$$\omega_b = [\, \varnothing \;\; \theta \;\; \Psi \,]^T \tag{3.4}$$

X, Y, and Z coordinates, which indicate the corresponding distances from the fixed frame of reference origin, are used to define the position of the vehicle. Usually, the coordinates are aligned as follows: down, east, and north. The vehicle's orientation is indicated by the angles σ, φ, and ψ. Using an equation of motion, the quadcopter's motion is explained i six degrees of freedom. This idea provides the vehicle's three-dimensional (3D) position, attitude, and orientation. The dynamic equation of the vehicle describes the variation of orientation and position with time.

1) Euler Angles

There are 3 Euler angles to take into account while analysing the motion of an item, especially a quadrotor: yaw, pitch, and roll. Pitch denotes rotation about the y-axis, roll denotes rotation about the x-axis, and yaw denotes angle about the z-axis. In addition, Vb stands for linear velocity, while the x, y, and z axes are represented by the values of u, v, and w, respectively. In equation (4), where ωb is the angular velocity, Ψ is the yaw rate of change, θ is the roll angle's change, and in this case, Ø is the pitch angle's change. The combination of the, gyroscopic torque vector, and gravitational vector and moment vector can be used by the quadrotor's dynamical equations to express the total nonlinear dynamics, as per the Newton-Euler formalism. Equation (3.6) indicates while the force required to accelerate the drone's mass ($m\dot{V}b$) and centrifugal force ($\omega b(m\dot{V}b)$) are combined, the total thrust from the propellers (Tb) and the gravity (gRT) are in equilibrium.

The mass of the drone is represented by m in this equation, along with its linear acceleration ($\dot{V}b$), angular velocity (ωb), linear velocity (Vb), force of gravity acting on the drone (g), rotation matrix (RT) from the body frame to the initial frame (in equation 3.9), and thrust from the rotors (Tb). Using Ib as the inertia matrix in the equation (3.12) and τb as the torque produced by the gyroscopic forces acting on the UAV body's rigid frame, equation (3.6) can be used to calculate the drone's angular acceleration. The rotation matrix between the inertial frame and the body frame is defined by equations (3.8) and (3.9).

Numerous techniques, including quaternions, Euler angles, and trigonometric functions, can be used to ascertain a vehicle's orientation. The simple Euler angle approach uses a transformation matrix to determine the vehicle's orientation related to the body frame. Due to the rotation about one of the body frames axes, the other two body frame axes are moved away from the inertial frame axes, assuming that the inertial frame of reference X, Y, and Z and the body frame of reference x, y, and z are the same. In the meantime, the rotation axis stays parallel to the corresponding inertial axis. Using equations for inertial trigonometric functions, the rotational body frame axes can be illustrated. The trigonometric equations can then be expressed using matrices. Two more matrices can be used to rotate the other two body frame axes. Figure 3.15 shows the body frame's x, y, and z axes rotation with respect to the corresponding axes of the inertial frame, as well as the equations for inertial trigonometric functions and their matrix representations. The transformation matrix from the inertial to the body frame is obtained by combining these rotational matrices. It matters what order things are done in. A rotation matrix is used in a quadcopter to convert from the inertial frame to the body frame. Wη and Wη-1 are the transformation matrices for angular velocities, respectively, from the body frame to the inertial frame and from the inertial frame to the body frame. ψ, θ, and φ are the three Euler angles that should be rotated in order to move from the earth reference frame to the body-fixed reference frame. To obtain the intended vehicle orientation, these rotations must be carried out in a certain order because they are not commutative.

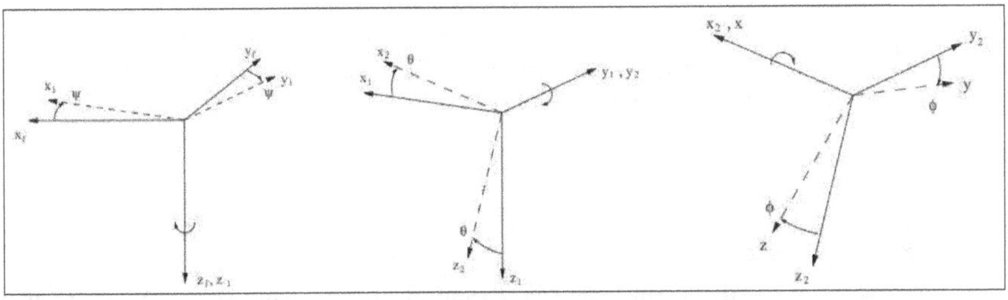

Figure 3.15 Frame transformation

$$\begin{bmatrix} b1 \\ b2 \\ b3 \end{bmatrix} = \begin{bmatrix} 1 & 0 & 0 \\ 0 & \cos(\Phi) & \sin(\Phi) \\ 0 & -\sin(\Phi) & \cos(\Phi) \end{bmatrix} \begin{bmatrix} \cos(\theta) & 0 & -\sin(\theta) \\ 0 & 1 & 0 \\ \sin(\theta) & 0 & \cos(\theta) \end{bmatrix} \begin{bmatrix} \cos(\Psi) & \sin(\Psi) & 0 \\ -\sin(\Psi) & \cos(\Psi) & 0 \\ 0 & 0 & 1 \end{bmatrix} \begin{bmatrix} x \\ y \\ z \end{bmatrix} \tag{3.5}$$

$$m_b \times \omega_b(mV_b) = gR^T + T_b \tag{3.6}$$

$$J^b\, \omega_b = -(\omega_b \times J^b\, \omega_b) + Tb - \tau b \tag{3.7}$$

$$R = R(\theta, z)\, R(\theta, y)\, R(\varphi, x) \tag{3.8}$$

R=

$$\begin{bmatrix} \cos\theta\cos\Psi & \sin\Phi\sin\theta\cos\Psi - \cos\Phi\sin\Psi & \cos\Phi\sin\theta\cos\Psi + \sin\Phi\sin\Psi \\ \cos\theta\sin\Psi & \sin\Phi\sin\theta\cos\Psi + \cos\Phi\sin\Psi & \cos\Phi\sin\theta\sin\Psi - \sin\Phi\cos\Psi \\ -\sin\theta & \sin\Phi\cos\theta & \cos\Phi\cos\theta \end{bmatrix} \tag{3.9}$$

The trigonometric equations are employed to generate three matrices, which are then multiplied to form a single matrix represented in Equation 3.10, which is the single transformation matrix needed to convert to the body frame from the inertial frame. The inverse transformation matrix is used to convert from the body frame to inertial frame. The transpose of Rt, or R1, is the inverse of R, since transformation matrices are orthonormal. The matrix from equation 3.10 is transposed in equation 3.11.

$$\begin{bmatrix} b1 \\ b2 \\ b3 \end{bmatrix} = \begin{bmatrix} \cos(\theta)\cos(\Psi) & \cos(\theta)\cos(\Psi) & -\sin(\theta) \\ -\cos(\Phi)\sin(\Psi)+\cos(\Psi)\sin(\theta)\sin(\Phi) & \cos(\Psi)\cos(\Phi)+\sin(\theta)\sin(\Phi)\sin(\Psi) & \cos(\theta)\sin(\Phi \\ \sin(\Psi)\sin(\Phi)+\cos(\Psi)\cos(\Phi)\sin(\theta) & -\sin(\Phi)\cos(\Psi)+\cos(\Phi)\sin(\Psi)\sin(\theta) & \cos(\theta)\cos(\Phi \end{bmatrix} \quad (3.10)$$

$$\begin{bmatrix} x \\ y \\ z \end{bmatrix} = \begin{bmatrix} cos\theta cos\Psi & -\cos(\Phi)\sin(\Psi)+\cos(\Psi)\sin(\theta)\sin(\Phi) & \sin(\Psi)\sin(\Phi)+\cos(\Psi)\cos(\Phi)\sin(\theta) \\ \cos\theta\, sin\,\Psi & \cos(\Psi)\cos(\Phi)+\sin(\theta)\sin(\Phi)\sin(\Psi) & -\sin(\Phi)\cos(\Psi)+\cos(\Phi)\sin(\Psi)\sin(\theta) \\ -sin\theta & \cos(\theta)\sin(\Phi) & \cos(\theta)\cos(\Phi) \end{bmatrix} \quad (3.11)$$

Equation 3.10 and 3.11 provide the orientation of the vehicle in relation to the body frame and inertial frame.

3.3.3 Equation of Motion

The crucial parameter required for the model has been identified, and the equation has been input into MATLAB Simulink for mathematical modelling and simulation. The resultant force acts perpendicular to the propeller shaft and applies pressure to the blade components. The thrust force is denoted is in equation (3.12). The quadcopter is manoeuvred to the anticipated position or state through the thrust generated by the propellers.

$$T = C_T \rho A(\omega R)^2 \quad (3.12)$$

$$M_Q = C_{MQ} \rho A(\omega R)^2 * R \quad (3.13)$$

$$M_R = C_{MR}\rho A(\omega R)^2 * R \qquad (3.14)$$

In equation (3.12), T represents the thrust force. It indirectly shows the effect of airflow between the propeller, with CT representing the thrust coefficient directly related to the rotor's thrust. In the equation, ρ stands for air density, R represents the rotor's radius, ω represents the rotor's angular velocity, and A represents the area of rotating the propeller. The torque, which is represented by the quadcopter's rolling moment MR and drag moment MQ, determines how much power the motor needs to maintain stability.

3.3.4 PID Controller

A controller is essential for the effective operation of any drone. Without a controller, a quadcopter is unstable. A PID controller consists of three control terms Proportional Integral and Derivative. The PID controller considers an error signal, which is the difference between the actual and desired value, as its input. It then multiplies this error signal by PID gains to control the plant or process. The proportional controller multiplies the error value by a constant gain value Kp. By integrating the error value over time until it approaches zero, the integral term eliminates the steady-state error. The derivative term first calculates the future error and then removes it.

3.3.5 Creating SIMULINK Model

In order to create a SIMULINK model which can accurately simulate the full dynamics of a drone, including gusty wind environments, the mathematical model of physics underlying the drone's operation is based on the TAROT FY650 frame. A PID controller is created and implemented for the drone, enabling that to precisely control its position and attitude. Figure 3.16 displays the PID controller along with drone model. Comparing the aforementioned model to the current multi-purpose models reveals advantages in terms of economy and serves as a more straightforward controller for a specific purpose. The parameters to construct the mathematical model are listed below in Table 3.4.

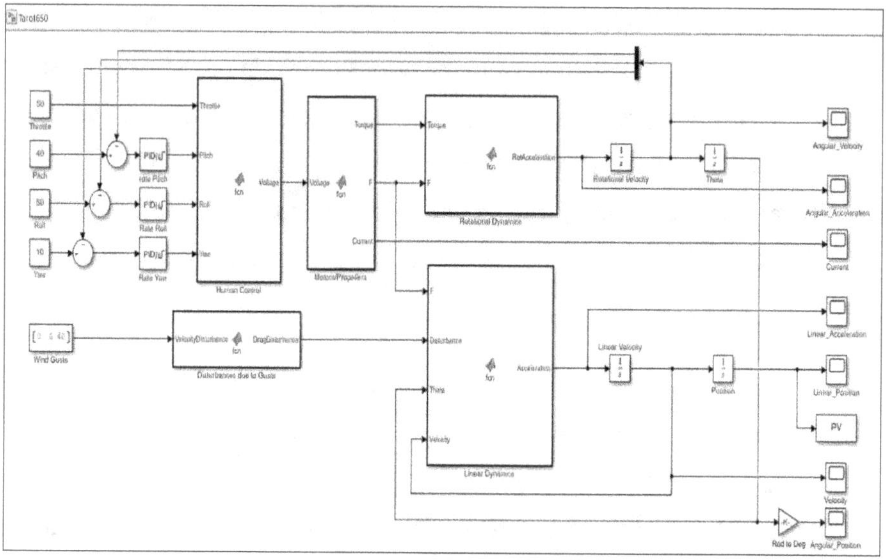

Figure 3.16. UAV Plant model with PID controller

Table 3.4 Tarot 650 Design Parameters

Diagonal length	335mm
Weight	1.780kg
Fontal Area	0.0197m²
Max Discharge Current	20C/77A
Motor KV	620 rpm/V
Propeller Size	13x5.5 inches

a) Battery Modelling

In a circuit, when components are connected in series, the overall voltage is the sum of the individual voltages of the components, while the current remains constant. On the other hand, when the components in a circuit are arranged in parallel, the circuit's voltage is constant regardless of the measurement location, and the current drawn is equal to the total current passing through all of its parts. Therefore, the following expression can be used to determine the nominal battery voltage: P = 3.8 * 6 = 22.8 P. The maximum energy that can be released from it is 191.5A at 20C, or I = 3.830 * 50.

b) Motor Modelling

The highest operational voltage of the battery is 22.8V, and its maximum operating current is 191.5A. These values can be used to calculate the average throughout a single battery discharge. These features make it possible to restrict the energy drawn by motors from the battery, ensuring that motor performance remains within a more practical range. All of these parameters will be included in SIMULINK once it is updated and properly implemented.

The KV, expressed in terms of RPM/Volt and equal to 620 in this instance, indicates the back-emf generated by the motor at any speed. To get the intended effects, the torque produced by the motor must be proportional to the square of the spinning velocity of the propeller. It is predicted that at 0V, RPM from KV would remain unchanged, and at the maximum voltage of 22.8V, RPM from KV will change by 25%. The following formula can be used to calculate the thrust in newtons using the coefficient of thrust.

$$Thrust = CT * \rho * n^2 * D^4 \tag{3.15}$$

The coefficient of thrust is denoted by C_T, air density at mean sea level (1.225 kg/m^3) is represented by ρ, n represents the revolutions per second, and D represents the propeller's diameter in meters. The thrust vector operates at a ninety-degree angle to the horizontal, perfectly perpendicular to the drone. The motor will receive the torque it needs from the propeller to overcome drag at the current rotational speed. It is necessary to extract the torque from the propeller data to quantify yaw and to estimate current consumption. In a DC motor, torque is directly proportional to current. This relationship can be expressed using the following formula: Where K_T represents the torque constant of the motor in Nm/Amps, and Torque = K_T * I, where I represents the current in Ampere. The torque constant KT of the motor can be expressed as KT = 1/KV = 1/620 = 0.000161, which is inversely proportional of the motor's RPM constant, KV.

c) Linear and Rotational Dynamics Modelling

It is presumed that the "X" configuration drone consists of two beams positioned perpendicularly to one another. It weighs 1.780 kilograms in total. It is expected that this mass is distributed uniformly throughout the drone. in order to predict a quadcopter rotational and translational motion in all three dimensions (X, Y, and Z), the motion of the object may be examined as it moves across space.

The force acts in the x direction are, $Fx = ma_x$ (3.16)

The force acts in the y direction are, Fy $= ma_y$ (3.17)

The force acts in the z direction are, Fz $= ma_z$ (3.18)

The similar kind of principles will apply to rotating motion,

The moment acts through the x-axis, also known as the pitch:

$$Mx = Ix * \ddot{\theta}x \tag{3.19}$$

The moment acts through the y-axis, also known as the roll:

$$My = Iy * \ddot{\theta}y \tag{3.20}$$

The moment acts through the z-axis, also known as yaw:

$$Mz = Iz * \ddot{\theta}z \tag{3.21}$$

where I is the moments of inertia in kilograms per meter square, M is the external moments or torques expressed in Nm, and rad/s2 is the rotational acceleration.

The moment of inertia of a rod can be calculated using the formula:

$$I=\int_0^M r^2\, dm \tag{3.22}$$

The moment of inertia of a point mass about a principal axis of rotation is given by:

$$I = mr^2 \tag{3.23}$$

The drone frame can be considered as four symmetrically positioned rods of equal mass around its centre of mass, which substantially simplifies the problem.

d) The Yaw, Roll and Pitch Moment of Inertia calculation

The moment of inertia of a rod can be calculated using the formula $I = 1/3\ ML2$, where m is the mass of the arm or arm in kilogram and L is arm length in meters. Since the drone is assumed to be perfectly symmetrical and appearances the same from all three sides, the pitch and roll moments of inertia are also the same. As a simple approximation, this can be applied as $I = 1/3\ ML^2$, where L is the distance from the centre of mass of the UAV while checking sideways.

e) Moments and Disturbances

The third law of Newton states that there is an equal and opposite reaction to every action. Each propeller's required torques have an impact on the UAV as well. Propellers 2 and 4 revolve in the opposite direction to the propellers 1 and 3. In each of these pairs, altering the power dynamic separately causes yaw.

$$M_z = T4 - T1 + T2 - T3 \quad (3.24)$$

f) Thrust and Moments

To induce pitch, F1 and F2 as well as F4 and F3 are oppositely coupled, similarly F4 and F1, are oppositely coupled with F3 and F2, to produce roll. The force of gravity acting on the drone's centre of mass is expressed as mg, where m is the mass of TAROT 650 (0.743 kg) and g is the gravitational acceleration (9.81 m/s^2).

g) Pitch and Roll

The drone pitches at an angle of θx as the forces on F3 and F4 rise. While F4 and F1 fall when rolling positively, F3 and F2 increase, causing θy to change but the total thrust to remain constant. The centre of mass is the path via which the gravitational force acts; it does not create a moment here.

$$M_x = (F3 + F4) * 0.237\,2 - (F1 + F2) * 0.2372 \quad (3.25)$$

$$M_y = (F3 + F2) * 0.237\,2 - (F4 + F1) * 0.2372 \quad (3.26)$$

h) Drag Disturbance

The estimated area A for the cube shape is 0.0197 m^2. The coefficient of drag (Cd) for a cuboid shape is 1.00. The area should be doubled because two beams are crossing each other. After accounting for the rotors, additional

30% to be added to address the interference drag that gives a total area of A = 0.0512 m^2. It's crucial to remember that the predicted area actually changes depending on the aircraft's pitch, roll, and flying direction.

$$Drag = 1/2 * \rho * V_{wind}^2 * A * C_d \quad (3.27)$$

3.3.6 Analysis of the Mathematical Model

The gathered data of a Tarot 650 quadcopter is presented in Table 1. The data was used in a simulation to closely match the physical characteristics of the quadcopter. Digital weighing scale is used to measure the mass of the frame, and the moment of inertia was measured using NX CAD. Practical restrictions govern the quadcopter's Simulink model's operation. The model was validated by applying motor speed inputs. Rather just charting the data in MATLAB, we achieved a more efficient and convenient visual representation of the results by using the Simulink scope block. In order to compare and track behavioural changes, the scope was utilized to plot psi, theta, phi and altitude

To produce thrust, each motor spins at the same speed at first. Pitch angle changes when motors M2 and M4 or M1 and M3 rpm are altered at the same time. In a similar vein, altering M2 and M1 or M4 and M3 speeds at the same time changes the roll angle. The four motors' individual speeds can be changed to alter the yaw angle. With altitude larger than zero, we varied the reference input values for pitch, roll, and yaw while maintaining zero for the other two parameters. The model performs as anticipated, according to this testing.

Applying the mathematical parameters to the mathematical model will take into account the drag force, disturbance, inertia, and actual position of the quadcopter. The output of the block to adjust the attitude controllers and altitude controller are also used. Simple gains can be used to create PID controllers using Simulink's built-in PID block. Integrated PID blocks were deployed to regulate the actual plant model values with respect to the intended input values, PIDs are available for pitch, roll, yaw, and altitude. The output of the PID controller is translated into motor commands through PWM for the taken X-configuration quadcopter.

CHAPTER 4

RESULTS AND DISCUSSION

The study gives the clear understanding and representation about the aerodynamic interaction between two propellers and its impact on the propeller performance. To evaluate the performance of the rotors, measurements of thrust, current, power, and efficiency are taken. The loadcell measurements are made in order to account for variations in rotor speed and varied spacings when comparing samples. The non-dimensional Coefficient of Thrust (C_T) and Coefficient of Torque (C_Q) form is obtained by converting the thrust and torque. The coefficients of thrust and torque are used to find the Figure of Merit (FM), a measurement of a rotor's efficiency (or power need) in proportion to the theoretical best derived using momentum theory. The comparison is done for the rotor performance in all conditions using FM. The equation needed to determine FM of the two-rotor setup is given by equation (1).

$$\text{Figure of Merit}_{(P1+P2)} = \frac{(C_{TL}{}^{3/2} + C_{TR}{}^{3/2})/\sqrt{2}}{(C_{QL} + C_{QR})} \quad (4.1)$$

$$\text{Thrust Coefficient } C_T = \frac{T}{\rho n^2 D^4} \quad (4.2)$$

$$\text{Torque Coefficient } C_Q = \frac{Q}{\rho n^2 D^4} \quad (4.3)$$

The propeller efficiency can also be found using the below formula

$$\text{Propeller Efficiency } \eta = \frac{J}{2\pi} \frac{C_T}{C_Q} \quad (4.4)$$

Where,

C_{TL} is the thrust coefficient of the left rotor,

C_{TR} is the thrust coefficient of the right rotor,

C_{QL} is the torque coefficient of the left rotor

C_{QR} is the torque coefficient of the right rotor

C_T is the coefficient of thrust

C_Q is the coefficient of torque

T is the thrust produced in N

Q is the torque produced in N m

ρ is the density of air 1.22kg/m^3

n is the Rotation Speed in seconds (RPS)

D is the propeller diameter

J is the Advance ratio

$$\text{Advance Ratio } J = \frac{V_A}{nD} \quad (4.5)$$

V_A is the Advance speed in m/s

The above formulae are used to calculate the coefficient of thrust, coefficient torque and figure of merit.

4.1 LATERAL SPACING RESULTS

The performance of UAV propellers can be greatly impacted when they are arranged with a lateral axis shift, which is a horizontal offset of rotating plane with respect to one another. The LS variations followed here are LS 2.2 to LS 3.8. It majorly affects equally both the propellers when it comes to thrust generation, distribution, and the aerodynamic interactions between the propellers such as induced flow and vortex effects. Each propeller has an impact on the airflow surrounding the propellers next to it. The airflow from one propeller may "wash" into another if they are too near together, which could result in turbulence or change the pressure and velocity fields that each propeller encounters. When propellers rotate, tip vortices are created. A propeller shifted laterally may come into contact with the vortex that another propeller has shed; depending on the circumstances, this could have a beneficial or detrimental effect on the effective thrust.

4.1.1 Lateral Spacing Thrust

Figure 4.1 shows the variation in thrust generation of propeller 1 and 2 with respect to the rotating speed of the propeller from 4000 rpm to 8000 rpm with 1000 rpm interval. Propeller 1 (P1) is placed on the right and Propeller 2 (P2) is placed on the left. Both the experiment and CFD thrust values are shown and compared to each other. In general, thrust rises as RPM rises. This is because increased RPM causes the propeller blades to spin more quickly,

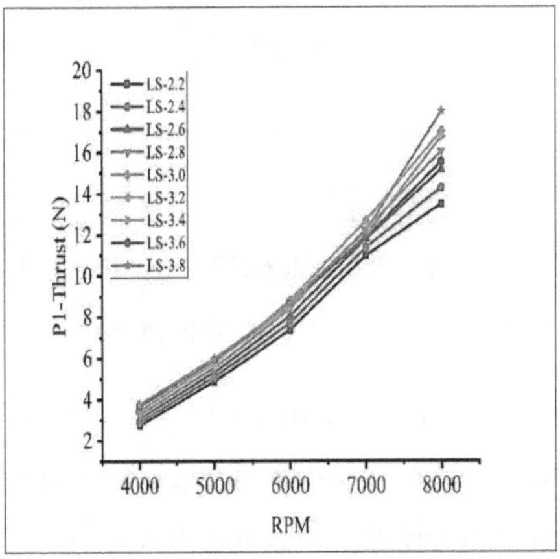

a) Propeller-P1 Thrust vs RPM

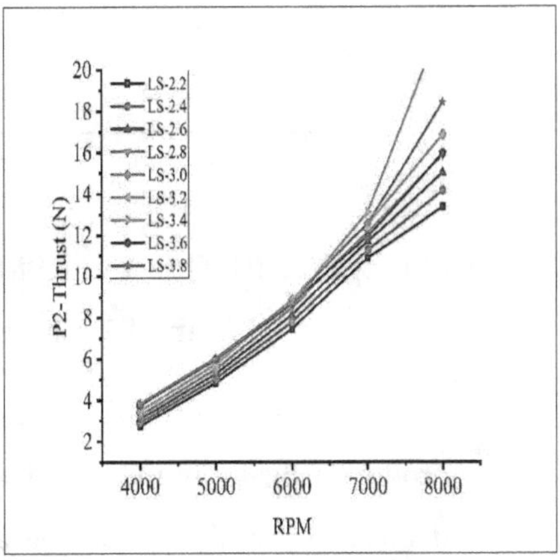

b) Propeller-P2 Thrust vs RPM

Figure 4.1 Experimental Thrust for Lateral Spacing

which pushes more air backward and produces more thrust. At lower RPMs, the relationship may be approximately linear, but as aerodynamic factors increase in significance at higher speeds, the relationship can often turn non-linear. From figure it is clear that at high rpm, the thrust variation is high particularly from the speed of 6000 rpm. This is mainly because of the aerodynamic losses and inefficiencies like drag and blade stall can cause reduced returns on thrust increase at higher RPMs as propeller tip speed approaches the speed of sound. For LS 3.6, LS 3.8, at higher rpm, fluctuations in thrust level were noticed. Considering the relationship between the thrust production and the lateral spacing, when the distance between the propeller is reduced laterally, the thrust is increased in both the propellers.

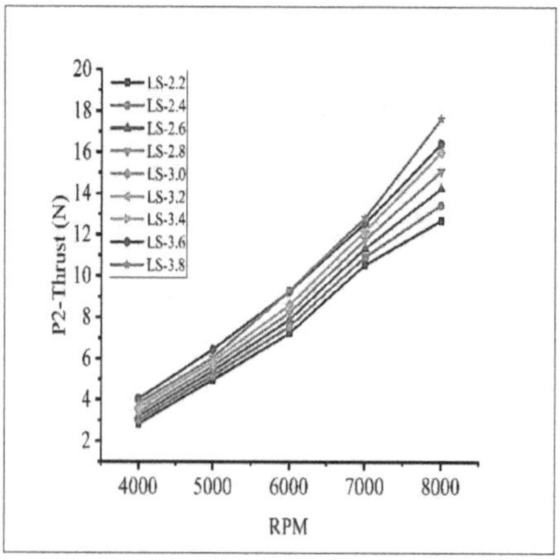

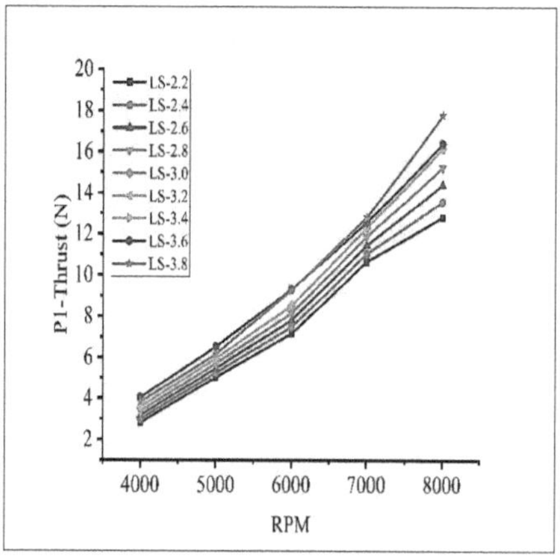

a) Propeller-P1 Thrust vs RPM b) Propeller-P2 Thrust vs RPM

Figure 4.2 Numerical Thrust for Lateral Spacing

From LS 2.2 to LS 3, the similar pattern was followed and the produced thrust is increased gradually. At lower spacing, the thrust level is decreased mainly due to the aerodynamic interference that arises when two propellers are in close proximity to one another. This includes complicated interaction between the airflow fields produced by each propeller. Due to turbulent wakes or disturbed airflow, these interactions may result in decreased efficiency, which lowers the thrust that is produced. However, at higher rpm, from 7000 rpm to 8000 rpm, the quantity of produced thrust is high and the thrust increment level is lower and almost the same for LS 3, LS 3.2, LS 3.4, LS 3.6 and LS 3.8. Once the propellers are separated and provided sufficient distance between them as LS 3, the interference effects on each other are nil and both the propellers are acting as a single propeller without any flow disturbance caused by the neighbour propeller as a result increased thrust is noticed which is higher than lower lateral spacing. The maximum level of thrust reached is 18.4N. At the lateral spacing 2.2 in the presence of higher vortices interaction, the produced thrust by the P1 and P2 are 2.724 N, 4.9 N, 7.382 N, 10.978 N, 13.493N and 2.751 N, 4.851 N, 7.456 N, 10.868 N, 13.358 N respectively. since both the propellers are in lateral axis and equally spaced, both propellers experience the equal amount of flow disturbances. The difference between the thrust level is as least as from 0.03 N to 0.15 N. The numerical results also seem the same compared to the experimental results with the values of 2.805 N, 5.022 N, 7.161 N, 10.648 N, 12.818 N and 2.833 N, 4.972 N, 7.232 N, 10.542 N, 12.690N for P1 and P2 respectively. Until it reaches the LS 3, the variation in thrust is 0.2 N to 0.3 N in each propeller but after LS 3, the increased thrust variation is started to decrease such as it is in the limit from 0.001 N to 0.01N that clearly indicates the reduced interference at higher lateral spacing from LS3. At LS 3, the thrust values are 3.467 N, 5.798 N, 8.78 N, 12.682 N, 17.018 N and 3.502 N, 5.740 N, 8.868 N, 12.555 N, 16.848 N for P1 and P2 respectively. After this LS, the minimal level thrust increase is notice in both the propellers. The numerical results shown in figure 4.2 also express the same pattern with the least variation compared to experimental tests.

4.1.2 Lateral Spacing Torque

For a number of reasons, torque measurement is essential to evaluating and improving UAV propeller performance. The effectiveness, stability, and functioning of UAVs in flight are directly impacted by an understanding of the capacity to control torque. Figure 4.3 shows the relationship between torque at various lateral spacing and at different speed from experimental tests. With respect to propeller rotating rpm, the torque value is increased at all lateral spacing.

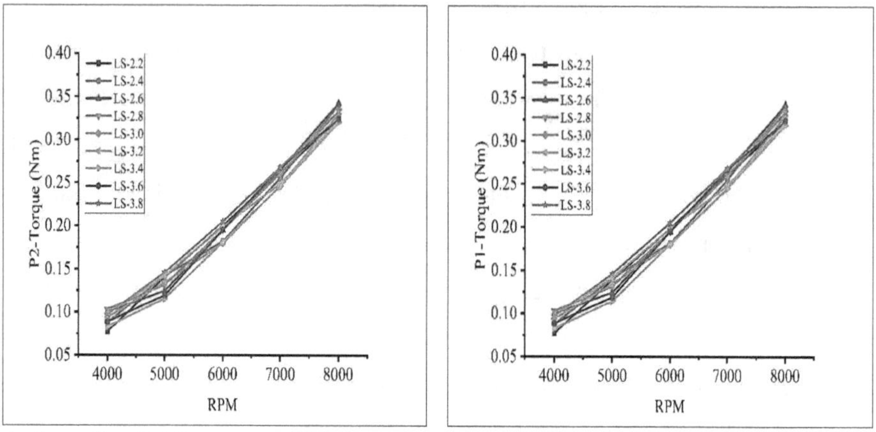

a) Propeller-P1 Torque vs RPM b) Propeller-P2 Torque vs RPM

Figure 4.3 Experimental Torque for Lateral Spacing

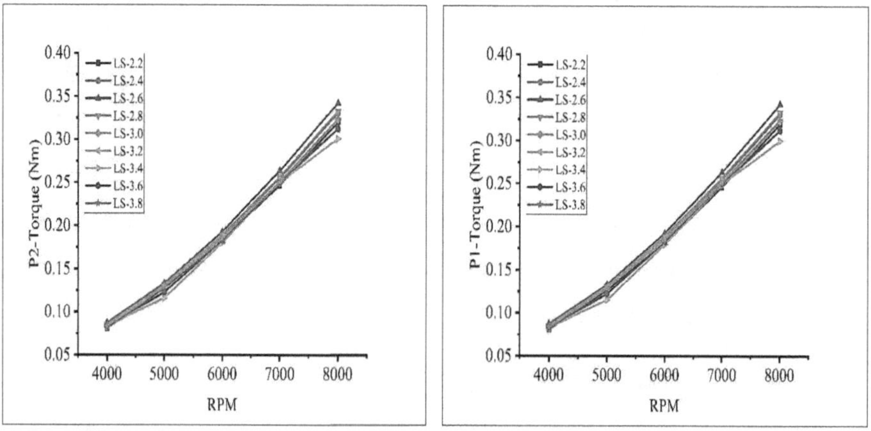

a) Propeller-P1 Torque vs RPM b) Propeller-P2 Torque vs RPM

Figure 4.4 Numerical Torque for Lateral Spacing

However, the variation noticed is not linear as it is in the thrust because of a number of mechanical, aerodynamic, and operational issues that might alter the propeller's performance. At lower lateral spacing, due to the effect of vortices interaction produced by both the propellers, random values are noticed. As the spacing is increased, the unstable flow started to settle down and the variation in torque is reduced which witnesses the reduced interference effect. LS 3.8 torque trend is linear with the values of 0.0973 Nm, 0.145 Nm, 0.205 Nm, 0.268 Nm, 0.333 Nm and 0.0971 Nm, 0.1458 Nm, 0.205 Nm, 0.268 Nm, 0.333 Nm for P1 and P2 propeller respectively. when the propellers are placed in extreme closer proximity as LS 2.2, 0.0769 Nm, 0.138 Nm, 0.198 Nm, 0.260 Nm, 0.323 Nm and 0.0772 Nm, 0.140 Nm, 0.200 Nm, 0.261 Nm, 0.324 Nm are produced at 4000 rpm, 5000 rpm, 6000 rpm, 7000 rpm and 8000 rpm respectively for individual propellers. From figures it is clearly understood that at lower spacing and at higher rpm, higher fluctuation in torque is observed. Because of the increasing centrifugal forces at higher RPMs, the propeller blades may bend, changing the aerodynamic forces acting on them as well as their effective pitch. The torque required is altered as a result. Depending on its position and rotational speed, each propeller blade encounters a distinct angle of attack during propeller rotation.

The lift and drag forces acting on the blade are influenced by the angle of attack, and this in turn impacts the torque needed to maintain a specific RPM. The amount of air displaced every revolution depends on the pitch, or the angle of the blades with respect to the incoming air. Variations in relative wind speed and angle cause each revolution to become less effective as RPM rises, changing the amount of torque needed. Compared to experimental results, numerical results shown in Figure 4.4 seem little linear. The reason behind this is that simulations frequently simplify the complex nature of aerodynamic interactions, such as the propeller-induced flow's effects on itself (propeller slipstream) and interactions. Variable environmental factors that can affect the experimental data, such as temperature fluctuations, humidity, and wind gusts, are often disregarded in numerical simulations.

Moreover, Numerical models usually exclude operational issues such as wear, vibrations, and mechanical errors. The CFD results for LS 2.2 P1 and P2 propeller are 0.08066 Nm, 0.12653 Nm, 0.18178 Nm, 0.24614 Nm, 0.31937 Nm and 0.08098 Nm, 0.12798 Nm, 0.18313 Nm, 0.24701 Nm, 0.32024 Nm respectively at the above mentioned rpm. For LS 3.8, 0.08335 Nm, 0.12962 Nm, 0.18590 Nm, 0.25114 Nm, 0.32427 Nm and 0.08322 Nm, 0.12950 Nm, 0.18561 Nm, 0.25078 Nm, 0.32427 Nm are the values in there which little variation only noted.

4.1.3 Lateral Spacing Thrust Coefficient

One of the most important factors in propeller analysis and design is the thrust coefficient. It offers a dimensionless way to evaluate a propeller's performance and efficiency, assisting in the characterization of the device's ability to convert rotational power into thrust.

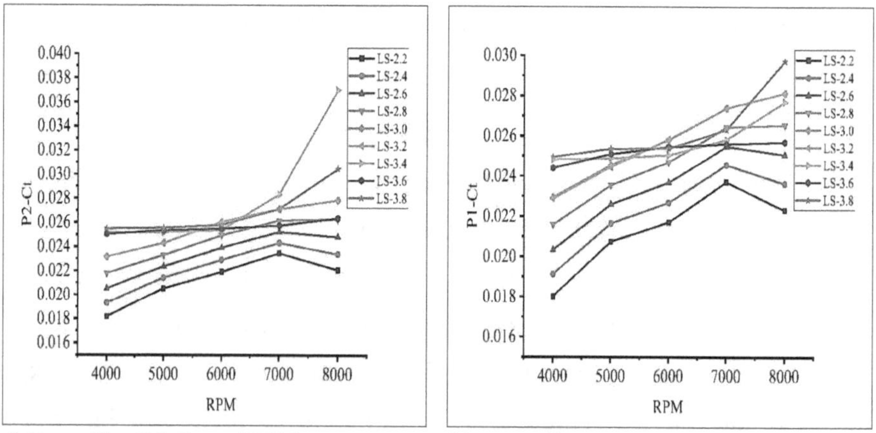

a) Propeller-P1 Ct vs RPM b) Propeller-P2 Ct vs RPM

Figure 4.5 Experimental Thrust co-efficient for Lateral Spacing

It offers a consistent method for evaluating the effectiveness of various propeller. Figure 4.5 and Figure 4.6 illustrates the thrust coefficient of propeller 1 and 2 at different rpm for both the experimental and CFD test results respectively. Thrust and thrust coefficient are directly proportional to each other. As such, the thrust coefficient is increased with rpm.

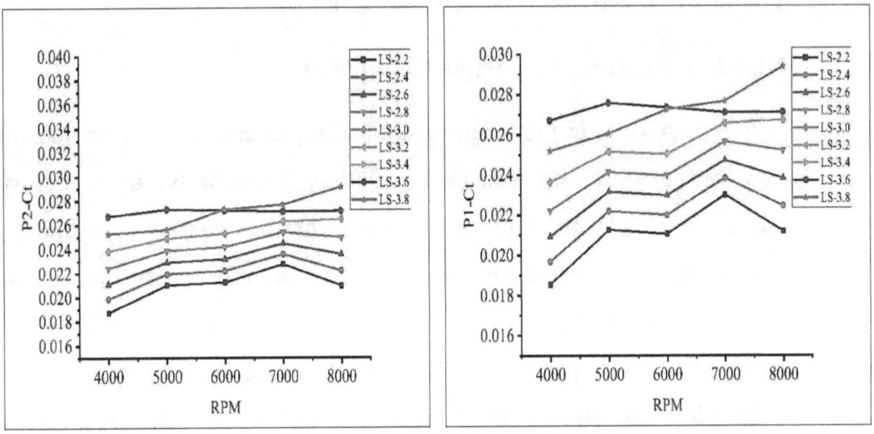

a) Propeller-P1 Ct vs RPM b) Propeller-P2 Ct vs RPM

Figure 4.6 Numerical Thrust co-efficient for Lateral Spacing

The trend is the same for thrust magnitude. At LS 2.2, the produced experimental test thrust coefficients are 0.01801, 0.02074, 0.02170, 0.02370, 0.02231 and 0.01819, 0.02053, 0.02191, 0.02347, 0.02208 for P1 and P2. While the obtained simulation values are 0.00399, 0.00400, 0.00399, 0.00397, 0.00395 and 0.01874, 0.02104, 0.02126, 0.02276, 0.02098 at the same LS 2.2 and at the early mentioned speed range. The LS 3 and LS 3.2 show better results at all rpm this is due to the optimal distance where the interference effects are minimal and used contructively for thrust increment. At LS 3.8, more unsteability is noticed and Ct values are reduced at higher rpm. from this also it could be understood that no interference also produce thrust coefficient decerement. The measured experimental values are 0.02495, 0.02536, 0.02538, 0.02632, 0.02971 and 0.02549, 0.02555, 0.02581, 0.02713, 0.03044 while CFD values are 0.02520, 0.02605, 0.02725, 0.02766, 0.02938 and 0.02528, 0.02563, 0.02730, 0.02767, 0.02915 for P1 and P2. Compred to all lateral spacings, LS 3 showed higher thrust coefficient with the values of 0.02293, 0.02454,

0.02581, 0.02738, 0.02813 and 0.02316, 0.02430, 0.02607, 0.02711, 0.02785 and the numerical values are 0.02362, 0.02515, 0.02503, 0.02656, 0.02673 and 0.02385, 0.02490, 0.02528, 0.02630, 0.02646.

4.1.4 Lateral Spacing Torque Coefficient

As it is in the torque graphical representation, the torque coefficient also varies with unpredicted fluctuation. Espectially, at higher rpm, and propellers are placed at a longer distance, grater variation in torque coefficient is noticed. Torque and torque coefficient are directly proportional to each other. The value variation in torque affects the torque coefficient. LS 3.8 showed the highest Cq at all rpm for both P1 and P2 propellers in the experimental and CFD plots as shown in Figure 4.7 and Figure 4.8 respectively. The corresponding values are 0.00481, 0.00462, 0.00451, 0.00433, 0.00412 and 0.00480, 0.00461, 0.00451, 0.00433, 0.00412 for experimental tests at the 4000 rpm, 5000 rpm, 6000 rpm, 7000 rpm and 8000 rpm respectively.

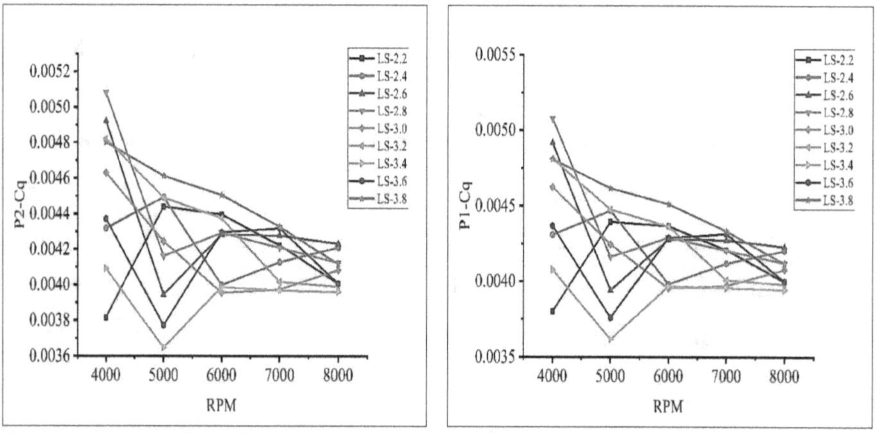

a) Propeller-P1 Cq vs RPM b) Propeller-P2 Cq vs RPM

Figure 4.7 Experimental Torque co-efficient for Lateral Spacing

The numerical results are 0.00412, 0.00410, 0.00408, 0.00405, 0.00401 and 0.00411, 0.00410, 0.00408, 0.00405, 0.00401 respectively.

Because both the propellers are placed in lateral direction, P1 and P2 propellers produced almost same results.

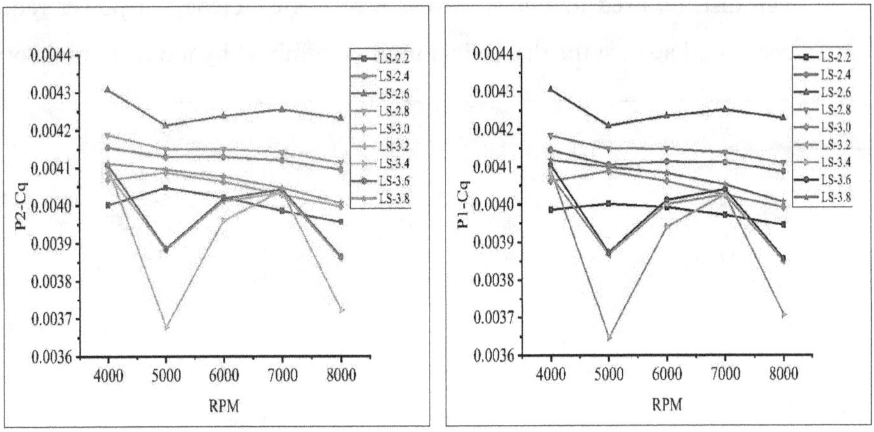

a) Propeller-P1 Cq vs RPM b) Propeller-P2 Cq vs RPM

Figure 4.8 Numerical Torque co-efficient for Lateral Spacing

The lowest Cq is noticed in LS 3.4 with the respective values of 0.00408, 0.00362, 0.00397, 0.00396, 0.00394 and 0.00409, 0.00365, 0.00399, 0.00397, 0.00396 in experimental tests. The numerical results are 0.00409, 0.00365, 0.00394, 0.00403, 0.00371 and 0.00411, 0.00368, 0.00396, 0.00404, 0.00372. The difference between the experimental and numerical results is only at the least level as 0.00481, 0.00462, 0.00451, 0.00433, 0.00412.The above graph illustrated that at lower lateral spacing, the torque coefficient insteability is reduced.

4.1.5 Lateral Spacing Figure of Merit

The Figure 4.10 shows the figure of merit (FM) at all lateral spacings taken for both the numerical and experimental tests. In the context of UAV propellers and rotor systems, the Figure of Merit (FoM) is an important efficiency parameter. It significantly affects the durability, and operational expenses of UAVs, making it especially important for their design and use. It is described as the ratio of the actual power used to hover to the ideal power needed to produce a certain amount of thrust. The thrust and torque coefficients can also be used to find it. FM is a rotor's efficiency or power requirement measured against the theoretical limit determined by momentum theory.

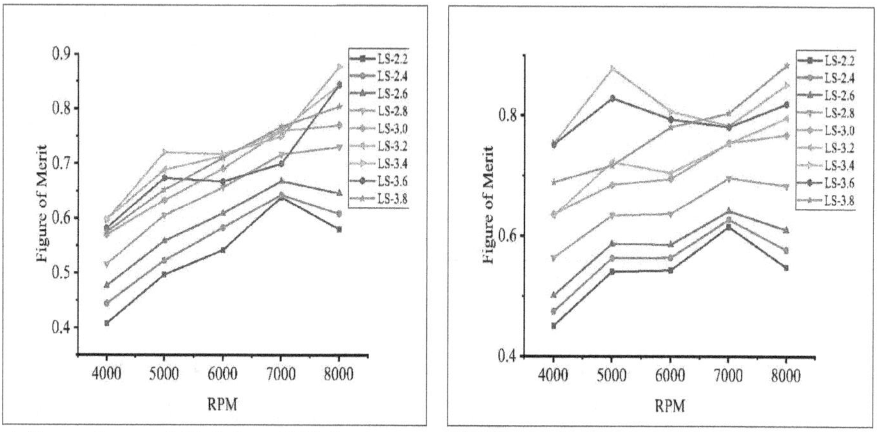

a) Experimental FoM vs RPM b) Numerical FoM vs RPM

Figure 4.9 Figure of Merit for Lateral Spacing

Compared to the all later spacing results, it is observed that as the rpm increases, the figure of merit is increased till the rpm of 7000, after which the FM is started to fall. This pattern is followed when the Lateral Spacing is lower to middle as from LS 2.2 to LS 3. This is because Propellers tend to run more effectively at lower RPMs because they produce less turbulent flow and aerodynamic drag. Since the actual power used is closer to the ideal power required to provide the same amount of thrust, this leads to a greater Figure of Merit. But an RPM that is too low might not produce enough force for steady flight. Higher RPM causes the propellers' tips to spin faster, which raises aerodynamic losses from tip vortices and increases drag. The propeller deviates from its ideal aerodynamic efficiency as a result of these influences, which lowers the Figure of Merit.

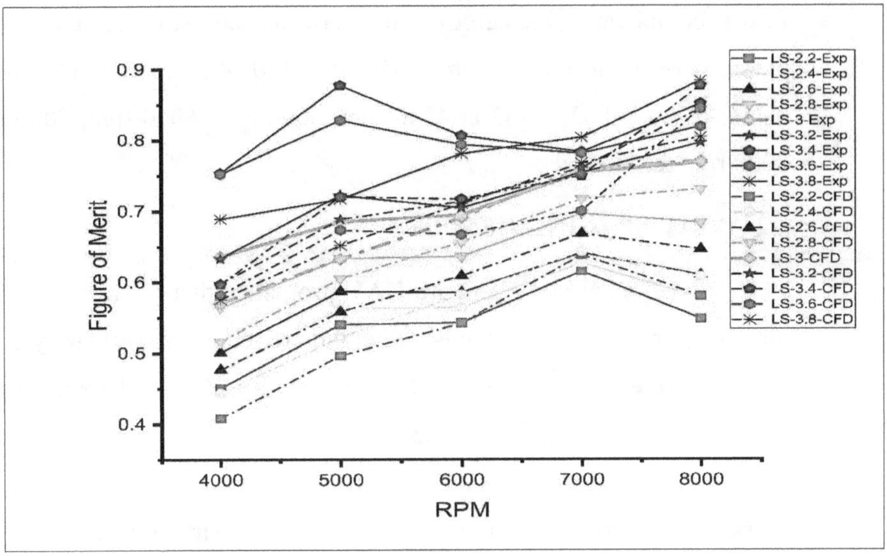

Figure 4.10 Comparative Figure of Merit for Lateral Spacing

Overall efficiency is reduced as a result of the higher power needed to compensate for these losses. At 7000 rpm, the FM for LS2.2, LS2.4, LS2.6, LS2.8, LS3, LS3.2, LS3.4, LS3.6 and LS3.8 are 0.642, 0.668, 0.716, 0.779, 0.760, 0.779, 0.759, 0.767 for numerical tests. The experimental test results are 0.615, 0.627, 0.641, 0.696, 0.754, 0.754, 0.782, 0.780, 0.803 for the same above mentioned respective lateral spacing in Figure 4.9. These are the highest FM for each test cases at 7000 rpm. At higher rpm, the figure of merit value is reduced with the values of 0.548, 0.576, 0.609, 0.683, 0.767, 0.795, 0.851, 0.818, 0.883 and 0.579, 0.607, 0.645, 0.730, 0.819, 0.843, 0.876, 0.844, 0.804 for both the experimental and numerical results. After the LS 3, the figure of merit values are not followed the linearity. The observed decrease in thrust may be attributed to the collapse of suction forces at the blade tips and an increase in vibrations due to flow separation. These phenomena tend to be relatively unstable at this particular rotor spacing, especially with a longer rotor arm. In such a configuration, the rotor is more likely to interact with its own wake, further complicating the aerodynamic performance. At LS 2.2 and LS 3.8, the obtained experimental FM are 0.450, 0.540, 0.542, 0.615, 0.548 and 0.689, 0.716, 0.780, 0.803, 0.883 at 4000 rpm, 5000 rpm, 6000 rpm, 7000 rpm and 8000 rpm respectively.

4.2 AXIAL SPACING RESULTS

Co-axial propellers are UAV propellers that are arranged one after the other along the vertical axis. Here, the propellers are shifted by 0.5, 1 and 1.5 R from the same central axis. There is a range of 0.5R to 1.5R in the distance between the rotating plane. Their interaction could potentially have a big impact on the performance on each propeller. The performance of each propeller is interdependent, and the performance variations are influenced by the close proximity of the two propellers. The disrupted airflow from the front propeller has a significant effect on the performance of the rear propeller, resulting in decreased thrust and efficiency.

4.2.1 Axial Spacing Thrust

Figure 4.11 shows the vertical shift thrust values at different horizontal spacing at different propeller rotating speed. compared to plain lateral spacing at different shift, each propeller is impacted differently. Propeller P1 which is placed behind the front propeller (P2) gets affected majorly by the induced flow by the propeller P2.The below figure compares the P1 and P2 thrust and also the experimental and numerical results. As the rpm increases, the amount of produced by each propeller is increased irrespective of the lateral spacing and axial spacing. This is due to the increased lift force, increased airflow and aerodynamic efficiency at higher rpm. The propeller blades' rate of air movement increases in correlation with the propeller's RPM. More air is displaced in a given amount of time as a result. In other words, at faster speeds, the propeller forces more air down. Airfoil wings and propeller blades have similar shapes. Each blade produces a greater lift force in proportion to its increased speed.

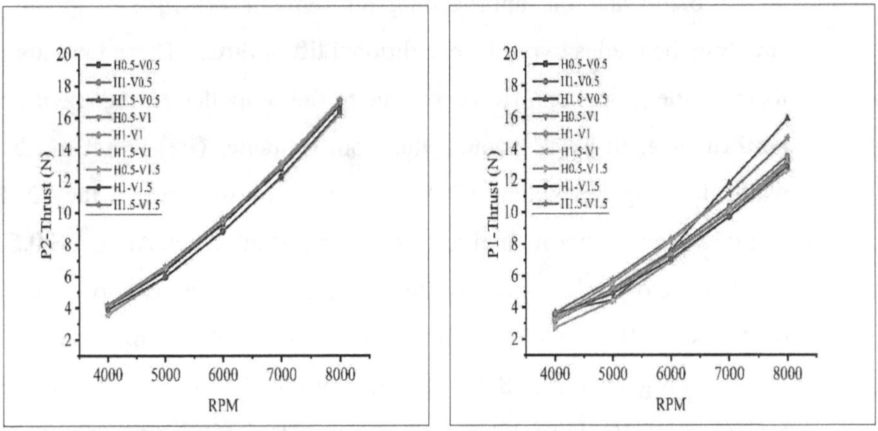

a) Propeller-P1 Thrust vs RPM b) Propeller-P2 Thrust vs RPM

Figure 4.11 Experimental Thrust for Axial Spacing

This lift is directed backwards in relation to the blade motion in a propeller, translating into a forward thrust. The propeller performs more effectively up to a certain speed.

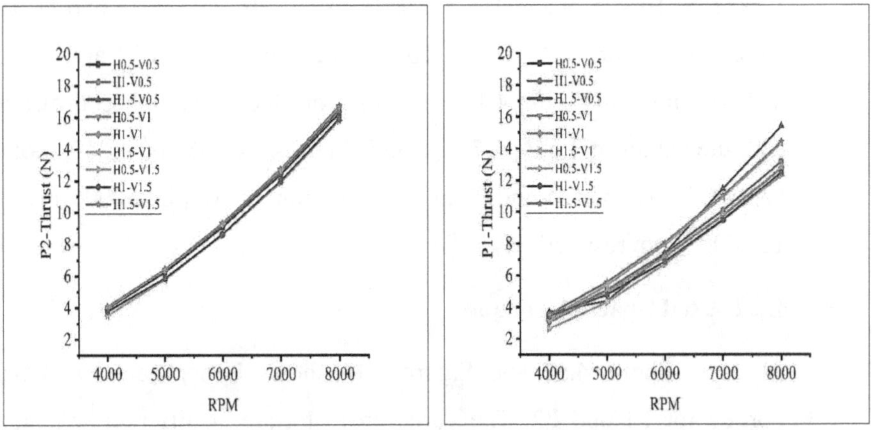

a) Propeller-P1 Thrust vs RPM b) Propeller-P2 Thrust vs RPM

Figure 4.12 Numerical Thrust for Axial Spacing

Before reaching the stall angle, where the airflow starts to separate from the blade surface, the angle of attack the angle formed by the chord line of the blade and the approaching air remains effective at greater speeds, enabling the blades to produce additional lift or thrust. These facts are common for both the propellers. However, due to the propeller arrangement affects the performance. In axial spacing, the front propeller (P2) is not much affected while the propeller behind (P1) is disturbed by the flow of the P2. Both the simulation and experimental results confirmed the same. At the LS 0.5 with VS 0.5, VS 1 and VS 1.5 showed reduced thrust where the tip vortices of P2 affect the thrust of P1. The VS is denoted as V and lateral spacing is denoted as H in graph. At higher rpm of 8000, the H0.5-V0.5, H1-V0.5, H1.5-V0.5, H0.5-V1, H1-V1, H1.5-V1, H0.5-V1.5, H1-V1.5 and H1.5-V1.5 experimental values are 13.192 N, 13.522 N, 15.885 N, 13.032 N, 13.112 N, 14.671 N, 12.605 N, 12.862 N, 14.655 N for P1 and 16.786, 17.024, 16.973, 17.038, 17.099, 16.205, 16.203, 16.376, 16.975 for P2 propeller. The respective simulation results are 12.771 N, 13.180 N, 15.408 N, 12.740 N, 12.793 N, 14.327 N, 12.298 N, 12.488 N, 14.425 N and 16.250 N, 16.593 N, 16.463 N, 16.656 N, 16.682 N, 15.825 N, 15.808 N, 15.900 N, 16.708 N respectively for P1 and P2 propellers that are shown in Figure 4.12 H1.5-V1.5 produced the highest thrust at all rpm with the values of 3.666, 5.764, 8.325, 11.184, 14.655 and 4.219, 6.656, 9.630, 12.949, 16.975 for P1 and P2 at the 4000 rpm, 5000 rpm, 6000 rpm, 7000 rpm and 8000 rpm respectively.

4.2.2 Axial Spacing Torque

Figure 4.13 and Figure 4.14 shows Experimental and Numerical Torque for P1 and P2. The variation in torque at different spacings, which shows the effect of rpm too. As it is in the thrusst, the same trend is followed in the torque values with respect to propeller rotating rpm in the rising flow. Compared to P1 and P2 propellers, P1 propeller torque values are not linear and much different from P2 front propeller. The forward propeller creates a complicated flow field that the aft propeller follows as it operates.

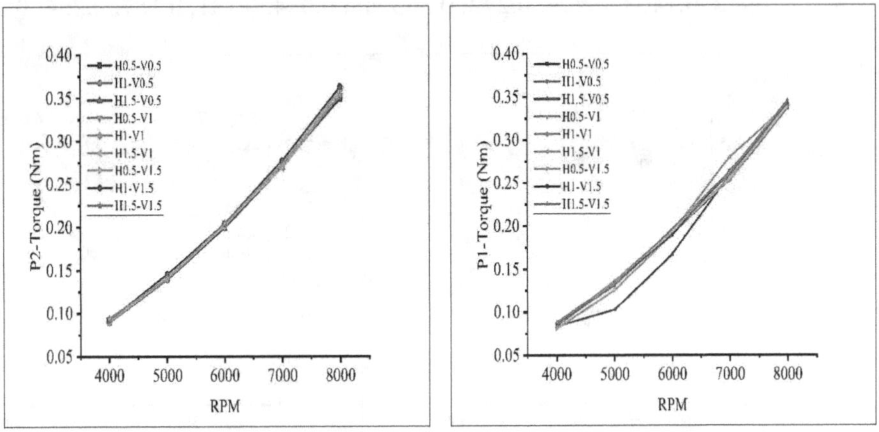

a) Propeller-P1 Torque vs RPM b) Propeller-P2 Torque vs RPM

Figure 4.13 Experimental Torque for Axial Spacing

The aft propeller's inflow conditions can be impacted by swirl, turbulence, and velocity deficiencies present in this wake. As the propeller blades travel through various wake zones, the disrupted inflow causes non-uniform loading on them, which results in torque fluctuations with each revolution. The rear propeller's blades may sometimes travel through areas with distinct wake characteristics left by the forward propeller's blades if the propellers' phases are out of sync. This may result in abrupt variations in the lift and drag forces acting on the aft propeller blades, which could then produce abrupt variations in torque. Vortices shed by the forward propeller remain downstream. The aft propeller's blades may encounter abrupt variations in force as a result of these vortices' interaction with the blades. The torque may spike or decrease each time a blade comes into contact with a vortex, which adds to the non-linear behavior of the vortex. 0.0835, 0.102, 0.167, 0.258, 0.343 and 0.0815, 0.098, 0.161, 0.244, 0.326 respectively for experimental and CFD tests of P1.

Unlike P1, here the gradual increment is noticed in P2 at all rpm and at all spacing that shows the less impact of the P1 propeller on P2. The results are 0.0942, 0.1403, 0.204, 0.272, 0.353 and 0.084, 0.131, 0.187, 0.254, 0.330.

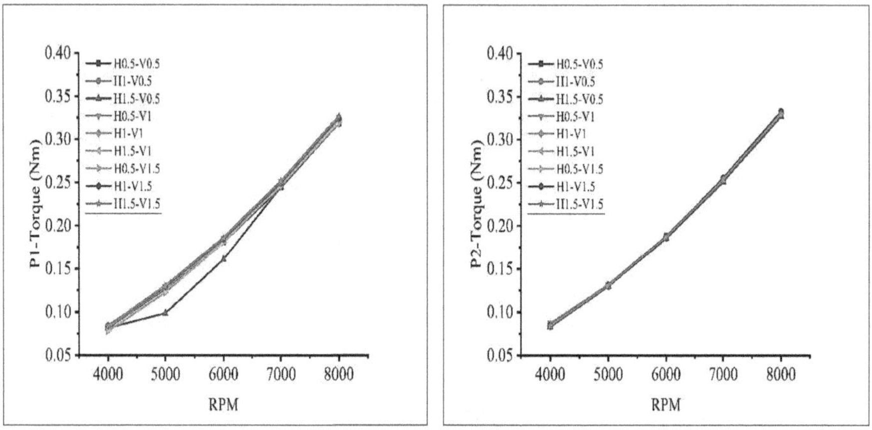

a) Propeller-P1 Torque vs RPM b) Propeller-P2 Torque vs RPM

Figure 4.14 Numerical Torque for Axial Spacing

4.2.3 Axial Spacing Thrust Coefficient

Figure 4.15 shows the relationship between thrust coefficient and propeller rotating rpm. As it is discussed previously, higher fluctuations are seen P1 aft propeller while P2 propeller values are little linear. However, the highest Ct is observed in both P1 and P2 at H1.5-V1.5 condition with the values of 0.0242, .02439, 0.0244, 0.0241, 0.0242 and 0.0279, 0.0281, 0.0283, 0.0279, 0.02806 respectively at the speed of 4000 rpm, 5000 rpm, 6000 rpm, 7000 rpm and 8000 rpm for experimental tests. The numerical results are 0.0234, 0.0235, 0.0236, 0.0237, 0.0238 and 0.0270, 0.0272, 0.0273, 0.0275, 0.0276 for P1 and P2 propellers. There will be less thrust produced by both propellers. Because of the asymmetric and disturbed inflow conditions, the thrust reduction of the front propeller will be less severe than that of the rear propeller. Since the back propeller runs in the wake of the front propeller, it will experience a very disturbed and non-uniform inflow.

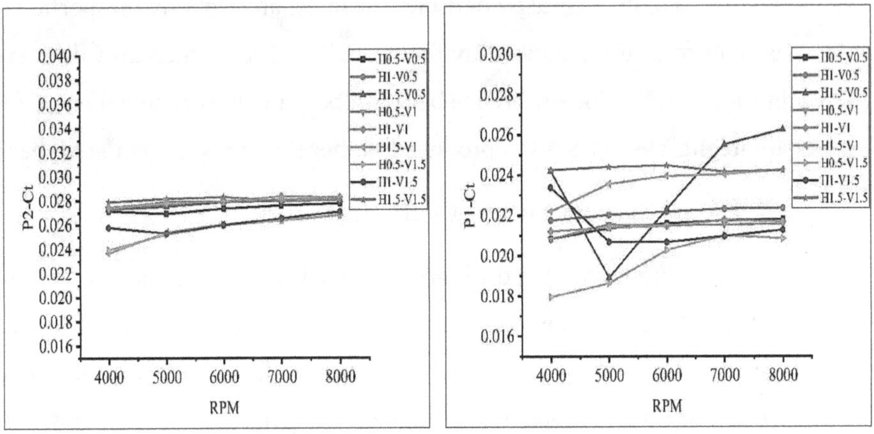

a) Propeller-P1 Ct vs RPM b) Propeller-P2 Ct vs RPM

Figure 4.15 Experimental Thrust Co-efficient for Axial Spacing

The thrust and general efficiency are greatly reduced as a result. The condition H1.5-V1 in P2 shows the lowest values with respect to rpm while H0.5-V1.5 provides the lowest in P1 propeller.

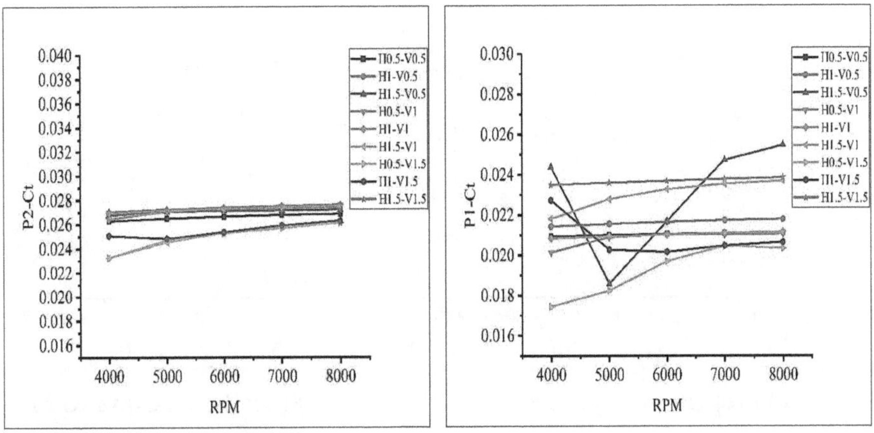

a) Propeller-P1 Ct vs RPM b) Propeller-P2 Ct vs RPM

Figure 4.16 Numerical Thrust Co-efficient for Axial Spacing

The above mentioned speed and spacing condition, the respective values are 0.0236, 0.0254, 0.0260, 0.0263, 0.0267 and 0.0179, 0.0186, 0.0202, 0.0209, 0.0208. Thrust coefficient values are directly proportional to the amount of thrust produced by the propeller. The Numerical values are shown in Figure 4.16. The difference between both the experimental and CFD values are negligible. H1.5-V0.5 produced higher thrust coefficient at higher rpm.

4.2.4 Axial Spacing Torque Coefficient

Figure 4.17 and Figure 4.18 shows the experimental and numerical results of the torque coefficient in relation to rpm and different lateral, horizontal spacings. It's a common known fact that the torque coeeficient results are varied under the impact of different reasons and conditions. Efficiency loss, uneven load acting on propeller and complex flow interactions are the main reasons for the varied Cq values. The system's overall efficiency will decline. This efficiency loss is caused by both the highly disturbed inflow for the rear propeller and the asymmetric stress on the front propeller.

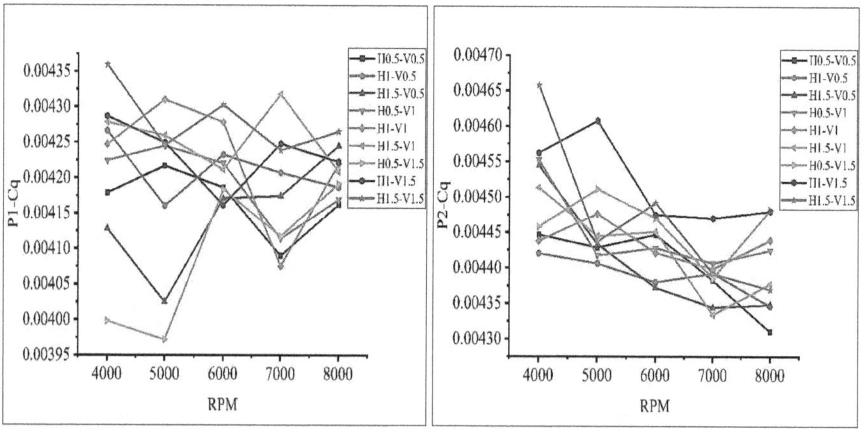

a) Propeller-P1 Cq vs RPM b) Propeller-P2 Cq vs RPM

Figure 4.17 Experimental Torque Co-efficient for Axial Spacing

In this setup, the flow interactions become quite complex. A difficult aerodynamic environment is created for both propellers by the wake from the rear propeller interacting with the front propeller in an uneven way.

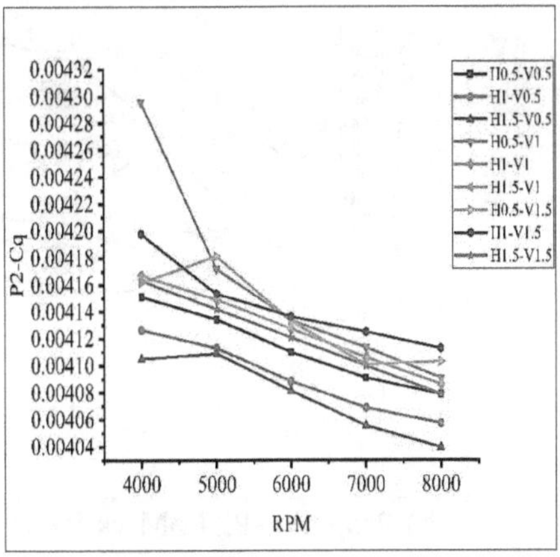

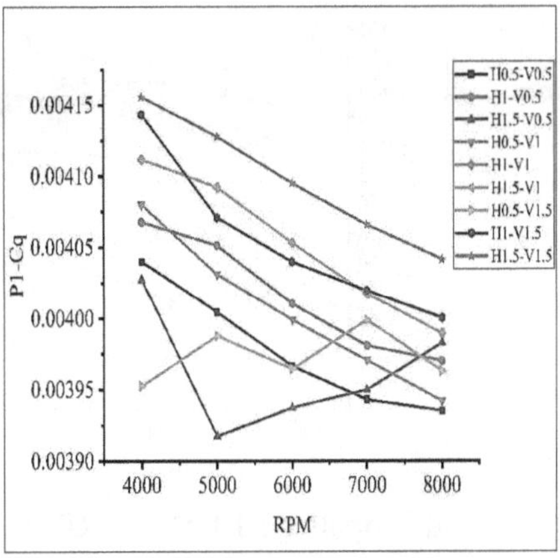

a) Propeller-P1 Cq vs RPM b) Propeller-P2 Cq vs RPM

Figure 4.18 Numerical Torque Co-efficient for Axial Spacing

For axial spacing, with respect to rpm, the produced torque coefficient values are reduced. H1.5-V1.5 showed higher Cq values in P2 while P1 H1.5-V1.5 showed higher values. The respective values are 0.00456, 0.00461, 0.00447, 0.00447, 0.00448 and 0.00436, 0.00425, 0.00430, 0.00424, 0.00427. The major fluctuations are shown in H0.5-V1.5 in P1 propeller and H1-V1.5 for P2 propeller. The below values are the obtained fluctuated values in the mentioned spacing as 0.00400, 0.00397, 0.00418, 0.00412, 0.00419 and 0.00456, 0.00461, 0.00447, 0.00447, 0.00448 from 4000 rpm to 8000 rpm with 1000 rpm interval.

4.2.5 Axial Spacing Figure of Merit

One of the most important metrics for assessing a propeller's efficiency is its Figure of Merit. In other words, it indicates the percentage of the power that is used to accelerate the air. With the distance increased, the figure of merit is also increased at all the tested rpm as a result of the decreased propeller flow interaction. The illustration of figure of merit for both experimental and numerical test are shown in Figure 4.19 and Figure 4.20.

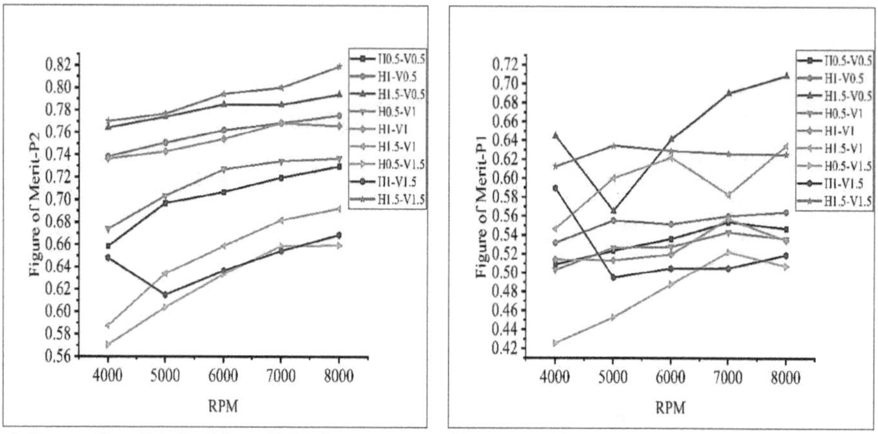

a) Propeller-P1 FoM vs RPM b) Propeller-P2 FoM vs RPM

Figure 4.19 Experimental Figure of Merit for Axial Spacing

Unlike lateral spacing, axial spacing requires the individual propeller performance measurement. For both the propellers, the figure of merit is calculated and moreover the combined effect figure of merit is also calculated.

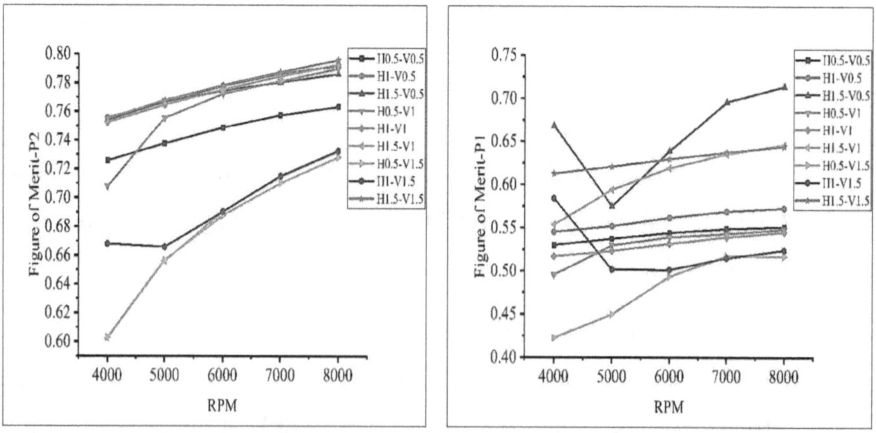

a) Propeller-P1 FoM vs RPM b) Propeller-P2 FoM vs RPM

Figure 4.20 Numerical Figure of Merit for Axial Spacing

From the individual propeller performance, it is vividly clear that the overall propeller efficiency figure of merit for P1 has major fluctuation and P2 has a least level of nonlinearity with respect to different spacing and also propeller rotating speed. Figure 4.22 represents the comparative FoM of both experimental and Numerical. A slipstream, or fast-moving column of air, is produced by the front propeller and affects the rear propeller. Because of the turbulent and uneven inflow, the rear propeller may perform less efficiently when operating in this disturbed airflow. Because it runs in undisturbed air, the front propeller often produces more thrust than the rear.

Each propeller will have a distinct power need. The rear propeller's performance is limited since it must deal with the front propeller's already-accelerated and disturbed air. Because of the less efficient working circumstances, the back propeller may require more power to produce the same amount of thrust as the front propeller.

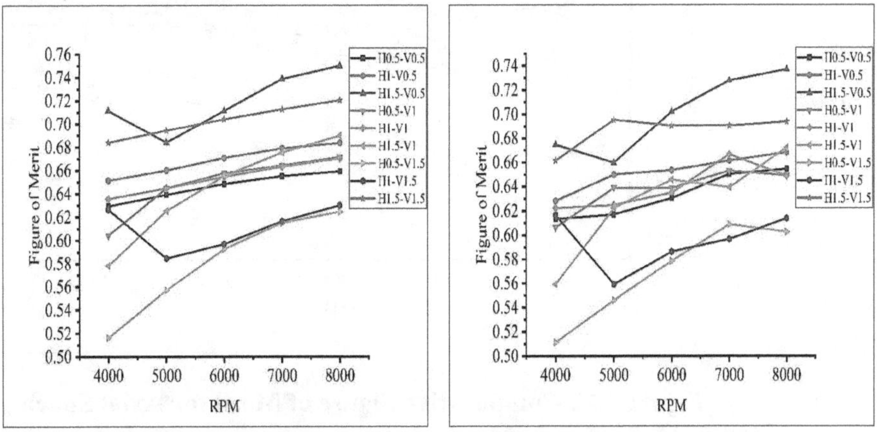

a) Experimental FoM vs RPM b) Numerical FoM vs RPM

Figure 4.21 Total Figure of Merit for Axial Spacing

The wake and vortices created by the front propeller may cause an uneven inflow into the back propeller. This may result in unequal loading on the back propeller's blades, which could raise mechanical stress and cause vibrations. At higher rpm of 7000 and 8000, H1.5-V0.5 shows the highest figure of merit for P1. At 8000 rpm, the experimental and CFD values are 0.5471, 0.564, 0.708, 0.536, 0.534, 0.634, 0.507, 0.519, 0.625 and 0.551, 0.572, 0.713, 0.548, 0.545, 0.646, 0.517, 0.524, 0.644 for H0.5-V0.5, H1-V0.5, H1.5-V0.5, H0.5-V1, H1-V1, H1.5-V1, H0.5-V1.5, H1-V1.5 and H1.5-V1.5 respectively. apart from these 2 rpm, the H1.5-V1.5 show the best performance. From 4000 rpm to 8000 rpm, the produced FM values are 0.612, 0.634, 0.628, 0.626, 0.625 and 0.612, 0.620, 0.629, 0.637, 0.644. For P2, different pattern is noticed.

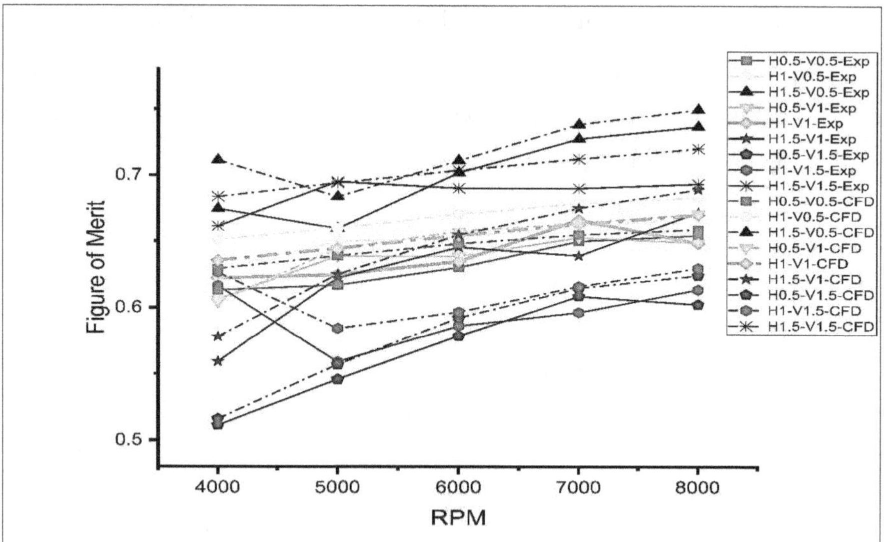

Figure 4.22 Comparative Figure of Merit for Axial Spacing

The pressure distribution surrounding the front propeller may slightly change when the rear propeller is present. This results from the two propellers' airflow and pressure fields interacting. But, in comparison to the effect of the front propeller on the back propeller, this effect is typically negligible.

Particularly, if the spacing is quite close together, the rear propeller may exert some back pressure on the front propeller. Although the effectiveness of the front propeller may be slightly reduced as a result, this effect is often much less significant than the opposite one. The noticed conditions are H1.5-V1, H0.5-V1.5, H1-V1.5. In these conditions, the front propeller is much affected by the presence of aft propeller. The higher impact is notived in H0.5-V1.5. The figure of merit values are 0.570, 0.603, 0.633, 0.658, 0.659 and 0.602, 0.656, 0.687, 0.709, 0.728 at the speed of 4000 rpm, 5000 rpm, 6000 rpm, 7000 rpm and 8000 rpm respectively. The highest FM is noticed in H1.5-V1.5 with the values of 0.661, 0.695, 0.690, 0.690, 0.693 and 0.683, 0.694, 0.704, 0.712, 0.720 in both the experimental and numerical tests respectively.

4.3 LATERAL SPACING CONTOURS AND PLOTS

4.3.1. Lateral Spacing Velocity Contours at 5000 rpm

The velocity contours that show the intensity of the flow and the interaction between the vortices are seen from Figure 4.23 to 4.26. When the velocity reaches 45 m/s at a 2.2 axis shift, the velocity distribution clearly illustrates the strong interactions between the two flows. The flow is clearly merging, resulting in a strong high-velocity core area. A noticeable vertical structure is formed between the streams as a result of the interaction, suggesting significant mixing and turbulence. While still interacting, the air jets are a little bit more apart at 2.6 than they were at 2.2 axis shift.

There appears to be less interaction intensity when the core high-velocity zone is narrower. While vortices are still evident, they are not as strong as those that were seen at the 2.2 axis shift. There is more control over the interaction. At 3, there is less core interaction and a greater sense of flow distinction. There is less merging if the high-velocity zones are concentrated closer to the rotor's lower side. Weaker and more dispersed are the vortices.

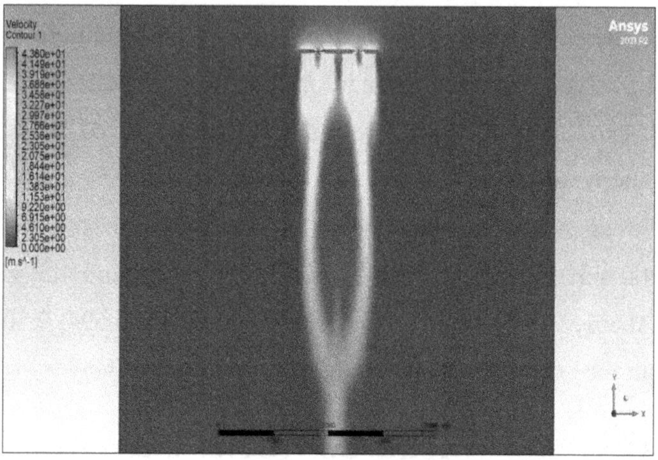

a) Downstream

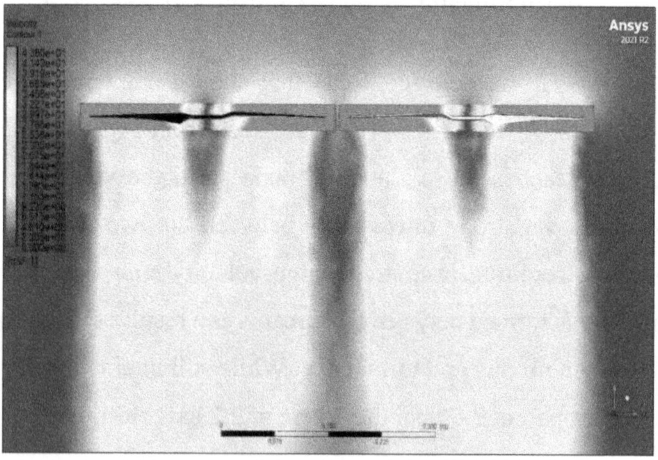

b) Close-up View

Figure 4.23 LS2.2 Velocity Contour at 5000 rpm

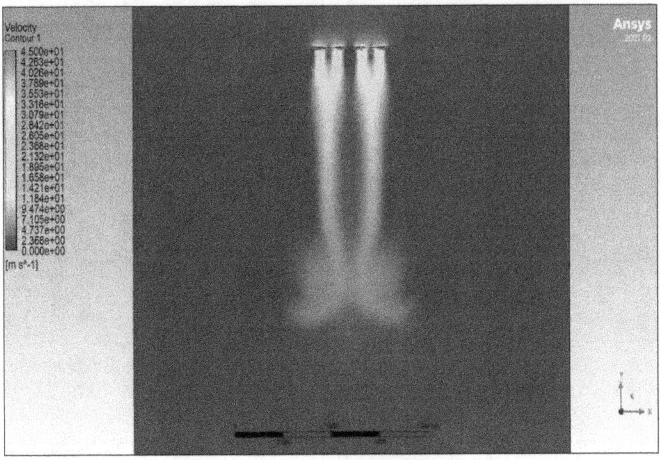

a) Downstream

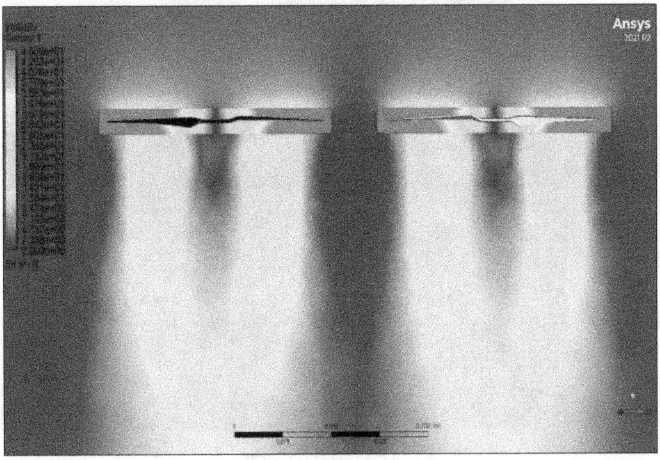

b) Close-up View

Figure 4.24 LS2.6 Velocity Contour at 5000 rpm

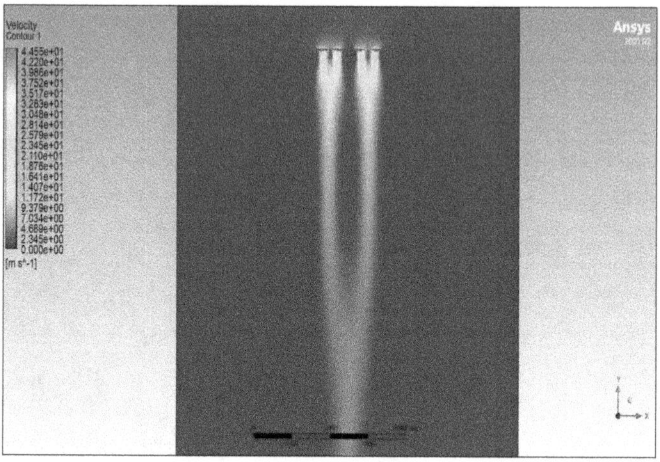

a) Downstream

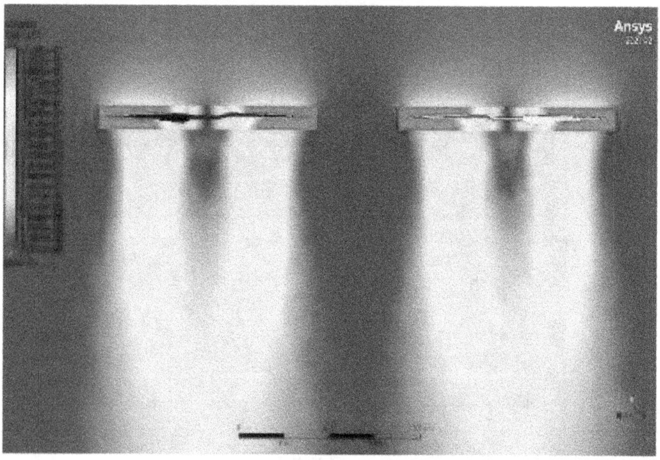

b) Close-up View

Figure 4.25 LS3.0 Velocity Contour at 5000 rpm

The flow pattern is more laminar because there is less turbulence between the jets. Similarly, at 3.4, there is very little interaction between the flows and they are almost totally independent. The localized high-velocity zones exhibit discrete streams of air. There is a notable reduction in the vertical structures. The flow is mostly laminar with little to no turbulence.

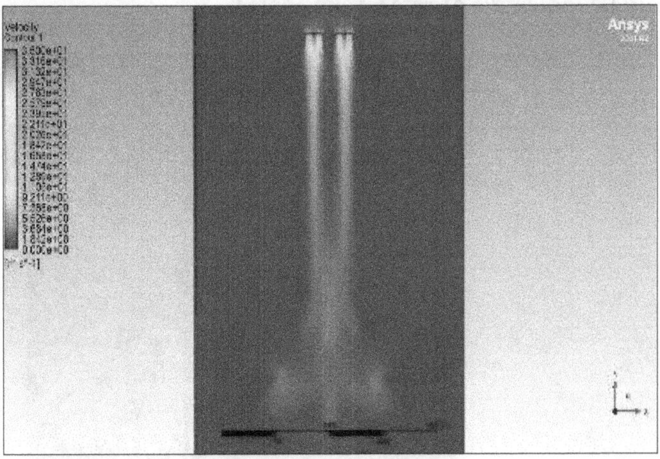

a) Downstream

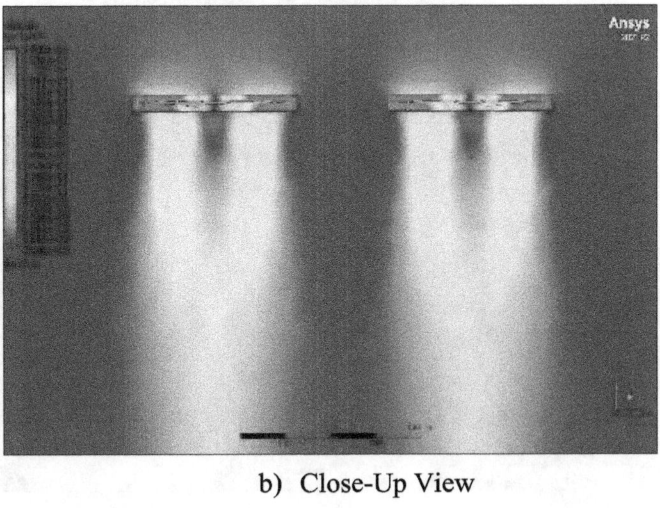

b) Close-Up View

Figure 4.26 LS3.4 velocity contour at 5000 rpm

4.3.2 Lateral Spacing Velocity Contours at 8000 rpm

Figures 4.27 to 30 display the velocity contours at 8000 rpm for lateral spacings of 2.2, 2.6, 3, and 3.4. It is evident from the contours that the velocity contours exhibit a more dispersed pattern with some abnormalities that could be indicative of rotor wake contact.

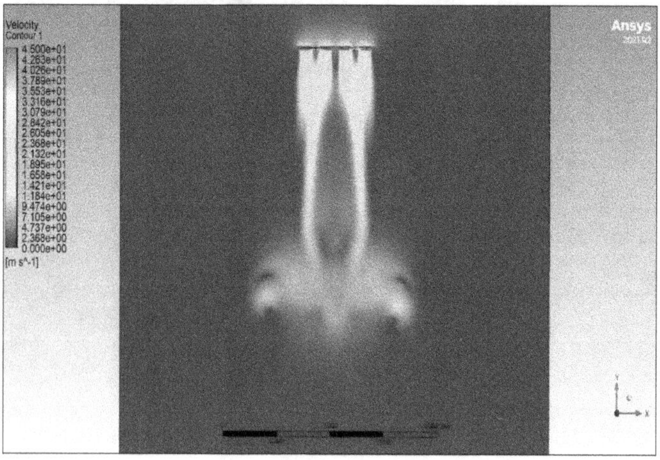

a) Downstream

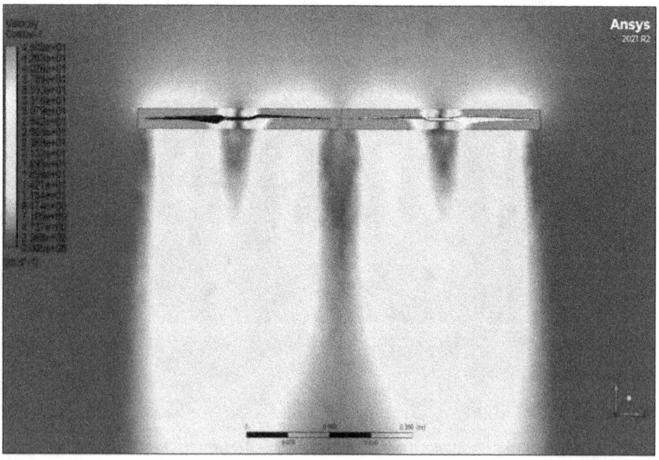

b) Close-Up View

Figure 4.27 LS2.2 velocity contour at 8000 rpm

a) Downstream

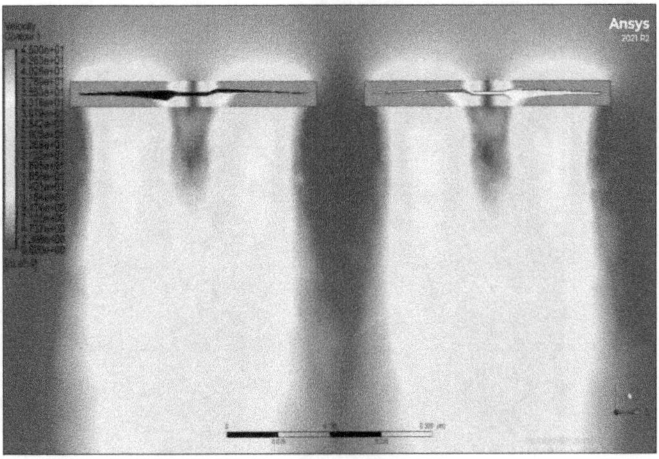

b) Close-Up View

Figure 4.28 LS2.6 velocity contour at 8000 rpm

The red areas indicate the increased velocity close to the rotors, which gradually decreases with increasing distance from the rotors. The speed curves seem more streamlined than the 2.2 axis shift. At the axis shift of 2.6, the rotor wakes interact less and the velocity distribution becomes more symmetric. The velocity contours in configuration 3 are the most streamlined of the three, all in the same progressive order.

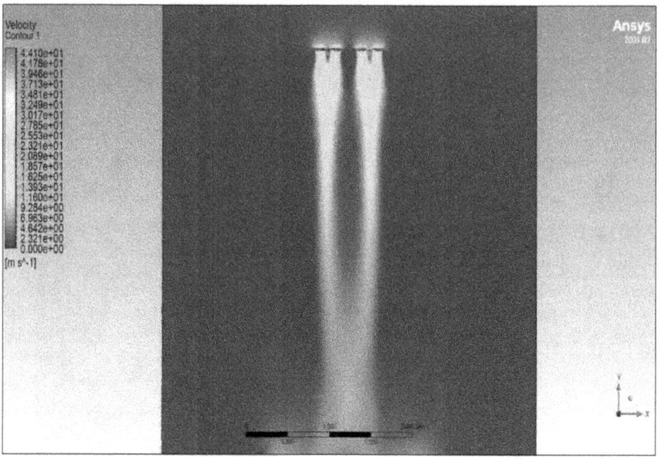

a) Downstream

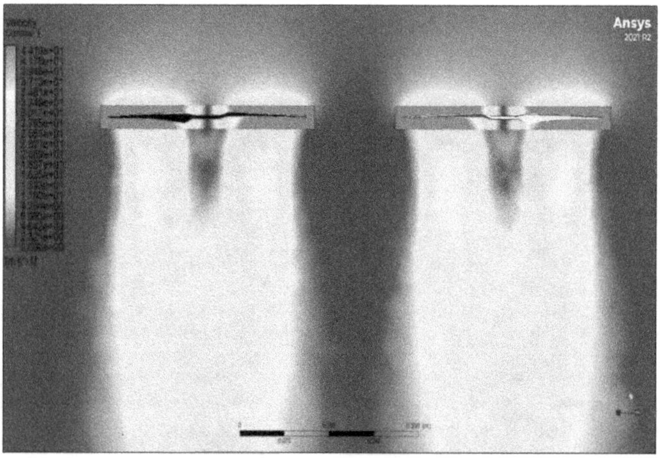

b) Close-Up View

Figure 4.29 LS3.0 velocity contour at 8000 rpm

Each rotor's wakes are clearly isolated from one another, suggesting less interference and a more effective thrust distribution. Due to less interaction effects, the flow is maintained with minimal losses after the axis shift 3. The interaction between the rotor wakes diminishes with increasing axis shift. Figure 4.29 shows more symmetric and elongated outlines, which suggest that a larger axis shift leads to less turbulence and wake interaction and a more effective thrust distribution. The 3.0 axis shift seems to be the best arrangement based on the contours, which could help the UAV operate better by minimizing wake interaction. Compared to 5000 rpm, the intensity of the vortices and turbulence are strong in the flow at 8000 rpm.

a) Downstream

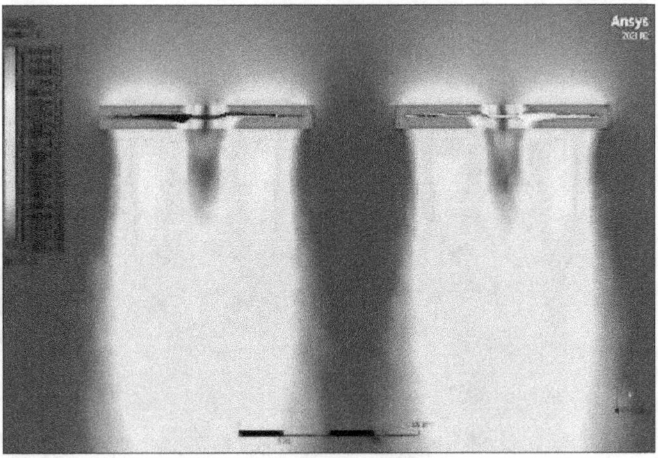

b) Close-Up View

Figure 4.30 LS3.4 velocity contour at 8000 rpm

The interaction between the flows diminishes with increasing axis shift. As a result, there is less mixing and turbulence. At greater axis shifts (3.0 and 3.4), the core high-velocity zone that is seen at smaller axis shifts (2.2 and 2.6) dramatically decreases. As the axis shift decreases, so do the vortices' strength and presence. Strong turbulence and vortices are seen at lower axis shifts, but the flow becomes more laminar and has weaker vertical features at greater axis shifts.

4.3.3 Lateral Spacing Velocity Streamline at 5000 rpm

Streamline contours show the flow variations in Figure 4.31 to Figure 4.34, lateral spacing ranges from 2.2 to 3.4 at 5000 rpm. Streamlines exhibit considerable vortex interactions and mixing at 2.2 axis shift, suggesting a strong aerodynamic coupling between the rotors. High levels of interference are indicated by the well-defined core region where streamlines converge. Significant rotor wake mixing results in noticeable interference effects. Higher aerodynamic interactions are suggested by the interaction zone's compactness and intensity.

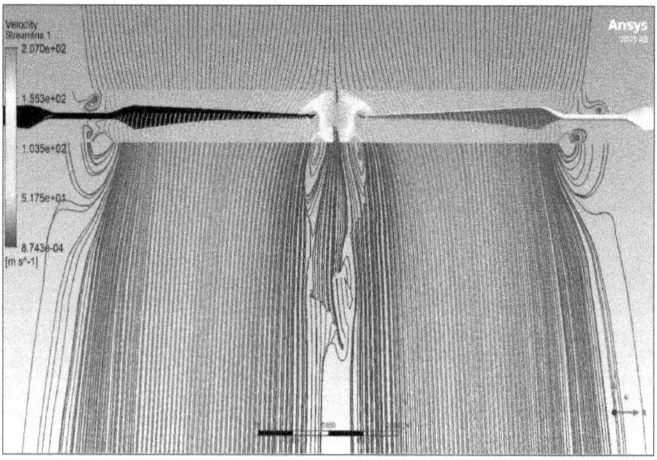

Figure 4.31 LS2.2 Velocity Streamline plot at 5000 rpm

There is a lot of interaction between the wakes from each rotor, which causes a lot of turbulence in the centre. Streamlines are more separated at 2.6 than they are at 2.2, which suggests less aerodynamic contact. Each rotor produces discrete, well-separated wakes, which reduce aerodynamic interference and turbulence.

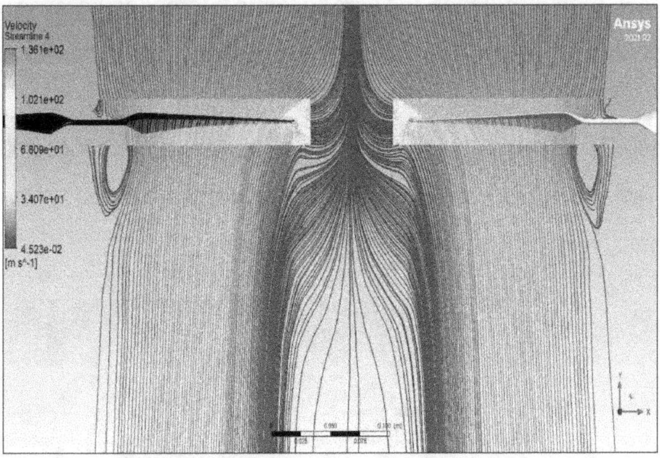

Figure 4.32 LS2.6 Velocity Streamline plot at 5000 rpm

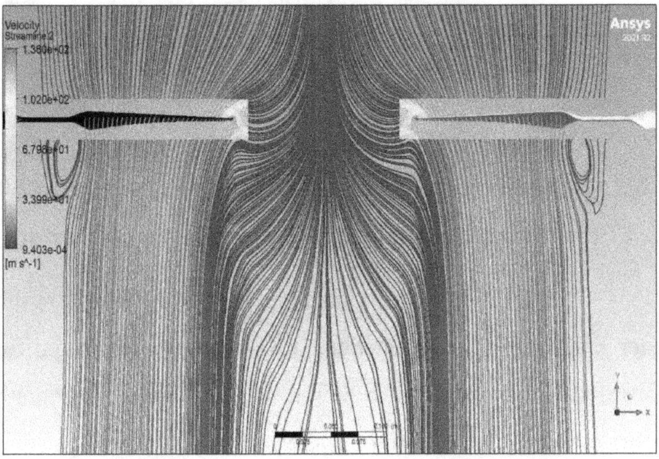

Figure 4.33 LS3.0 Velocity Streamline plot at 5000 rpm

The generated wakes are independent and do not interact closely as the distance between the rotors rises, reducing the adverse effect. Each rotor produces discrete, well-separated wakes, which reduce aerodynamic interference and turbulence. The generated wakes are autonomous and do not interact closely as the distance between the rotors rises, reducing the adverse effect. high turbulence and aerodynamic interaction because of the considerable rotor wake mixing. The wakes have a strong connection at axis shift 2.2, and there is modest interaction with less noticeable wake mixing at axis shift 2.6. There is less turbulence since the wakes are more pronounced. At 3 and after 3, there is much less interaction with clearly separated wakes, the flow is more stable, and there is much less turbulence.

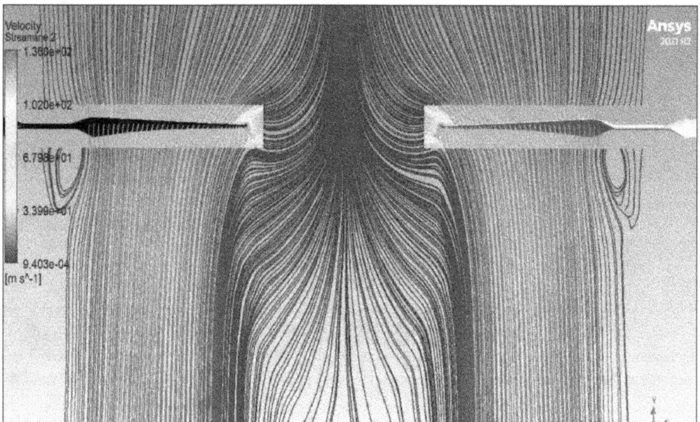

Figure 4.34 LS3.4 Velocity Streamline plot at 5000 rpm

4.3.4 Lateral Spacing Velocity Streamline for 8000 rpm

Higher RPM can provide more intricate streamline patterns and faster rotational speeds. Because of the higher centrifugal forces, you may see more noticeable changes in streamline curvature, stronger wake effects, and vortex interactions. The rotor wake and its surroundings can have substantially higher velocities at higher RPMs. Significant changes in the flow field and finer velocity contours may result from this. The rotor wake and its surroundings can have substantially higher velocities at higher RPMs. Increased turbulence, more dramatic fluctuations in the flow pattern, and possibly larger losses or performance problems can result from this.

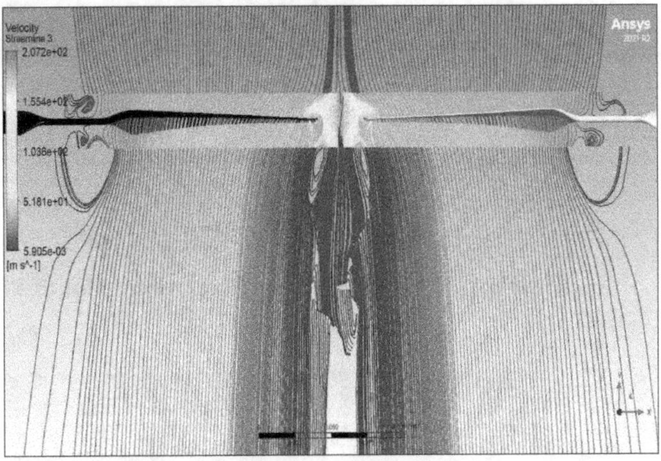

Figure 4.35 LS2.2 Velocity Streamline plot at 8000 rpm

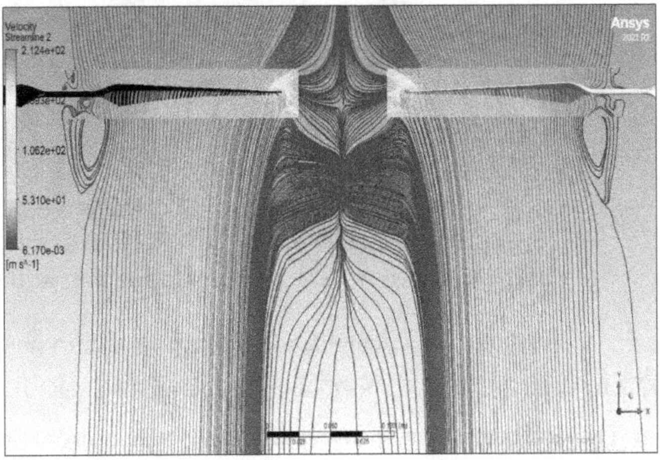

Figure 4.36 LS2.6 Velocity Streamline plot at 8000 rpm

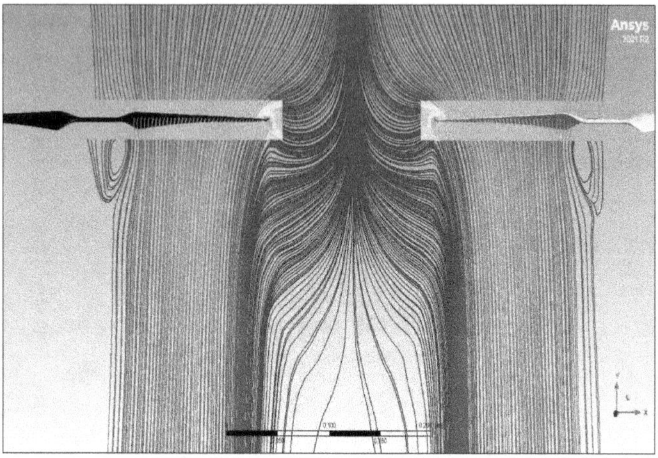

Figure 4.37 LS3.0 Velocity Streamline plot at 8000 rpm

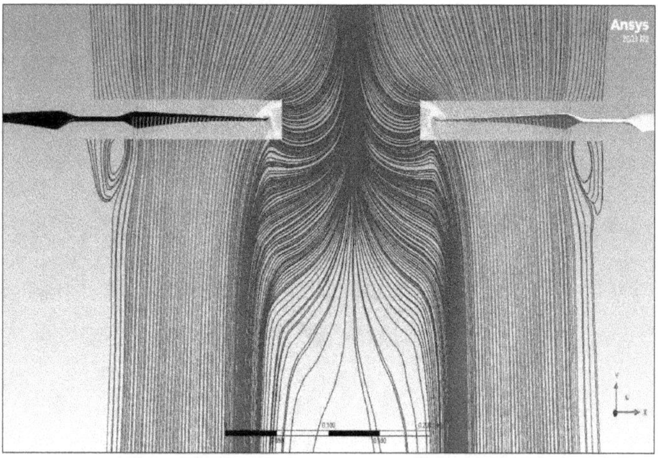

Figure 4.38 LS3.4 Velocity Streamline plot at 8000 rpm

For both LS 2.2 and LS 2.6, the maximum velocity range scale is 207.2 m/s and 202 m/s, respectively, exceeding the streamline velocity at 5000 rpm. The next rotor is more susceptible to turbulence from the other rotor when the rotors are closer together. Vibrations may rise and performance might drop as a result. The interference effects decrease as the distance increases from LS

2.2 to LS 3.4, and Figures 4.35, 4.36, 4.37, and 4.38 show stable, distinct stream lines. Sufficient distance minimizes turbulence's effect on the rotor and preserves flow stability.

4.3.5 Lateral Spacing Y-Velocity component at 5000 rpm

In Y-Velocity contours, the vortices interaction region can be seen clearly by interference region, velocity gradients, flow distribution and wake formation.

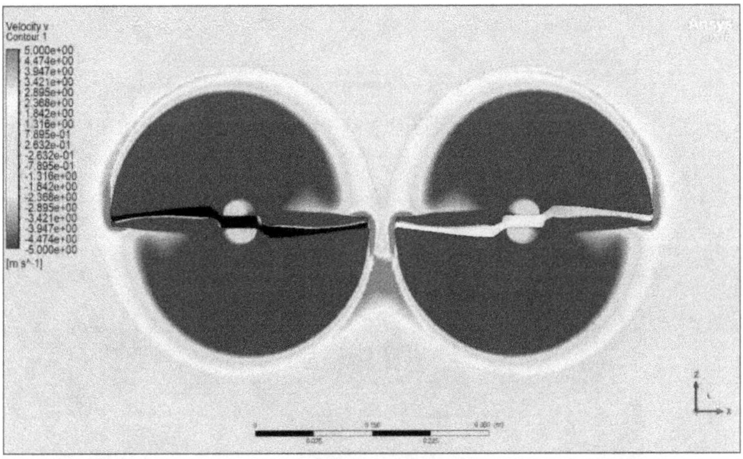

Figure 4.39 LS2.2 Y-Velocity Component at 5000 rpm

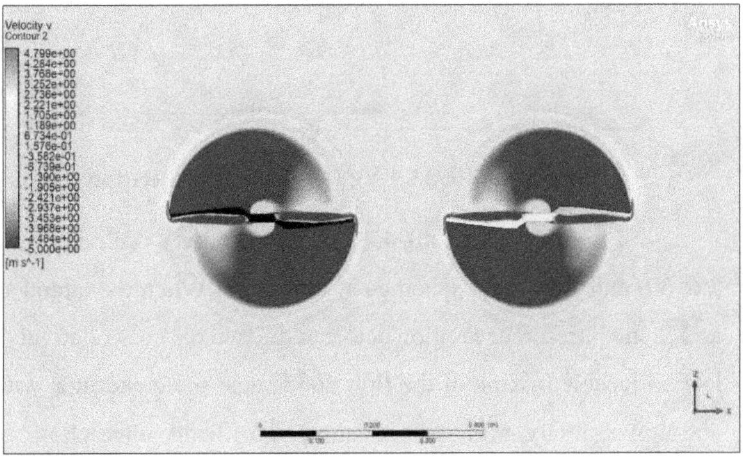

Figure 4.40 LS2.6 Y-Velocity Component at 5000 rpm

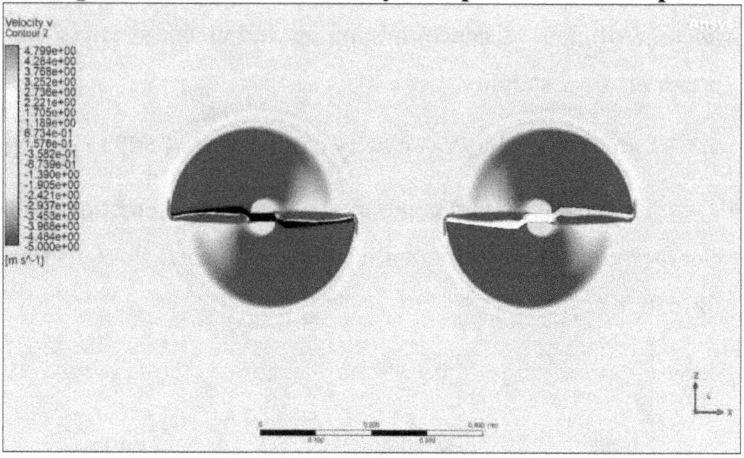

Figure 4.41 LS3.0 Y-Velocity component at 5000 rpm

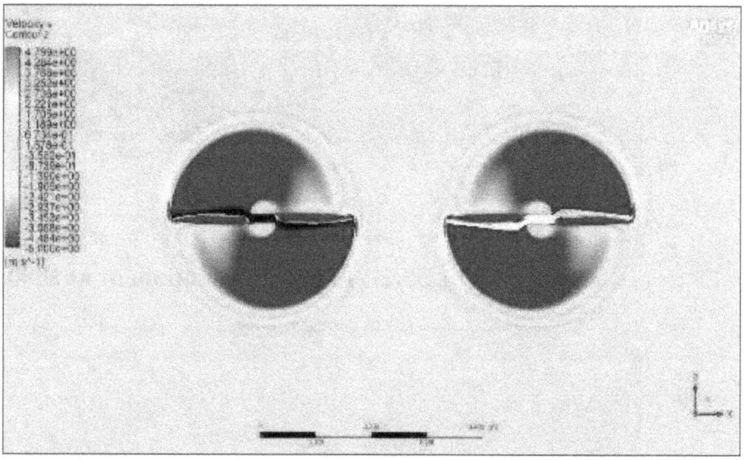

Figure 4.42 LS3.4 Y-Velocity component at 5000 rpm

Figures 4.39, 4.40, 4.41 and 4.42 show Y-Velocity component of 2.2, 2.6, 3.0 and 3.4 lateral spacings at 5000 rpm. When the lateral spacing is least as 2.2, the interference region between the two rotors is relatively narrow. There is a noticeable mixing of the flow fields near the centerline, with distinct high and low-velocity regions indicating significant interaction. Sharp velocity gradients are observed near the rotor tips and in the wake region. The high-velocity regions and low-velocity regions are more distinct, indicating strong interaction effects. The flow distribution appears more asymmetric, especially near the interference region. The uneven flow distribution could lead to potential performance issues and instability. The wake region behind the rotors shows significant disturbances and fluctuations. The wakes from each rotor interact closely, leading to complex flow patterns. At the lateral spacing 3, When considering the 2.2 axis shift, the interference region is wider. The middle region exhibits less prominent high and low-velocity areas due to the more dispersed mixing of the flow fields. There is more graduality and smoothness in the velocity gradients. In comparison to the 2.2 axis shift, the difference between the high and low velocity zones is less noticeable, indicating a decrease in interaction effects. The dispersion of the flow is more uniform and symmetric. Better aerodynamic performance may arise from a more balanced interaction between the rotors, as seen by this symmetry. There is less disruption and greater organization in the wake creation. Smoother flow patterns result from the wakes from each rotor being more distinct and less dependent on one another. This clearly shows that interaction effects are reduced.

4.3.6 Lateral Spacing Y-Velocity component at 8000 rpm

It is clear from the Y-Velocity contours that the same flow pattern and rotor contact are present at 8000 rpm compared to 5000 rpm, but with more intensity which are vividly seen in figures from 4.43 to 4.46 for the described four lateral spacing cases respectively. There is less maintained symmetry and greater degrees of turbulence and aerodynamic interactions at the lateral spacing of 2.2. The larger red and yellow regions indicate more aerodynamic contact and turbulence.

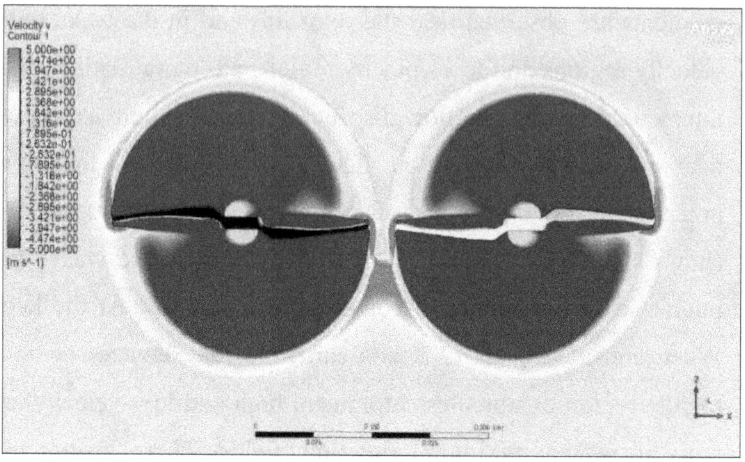

Figure 4.43 LS2.2 Y-Velocity Component at 8000 rpm

With a lateral spacing of 3, it is evident that the rotors' wake interactions with one another are less turbulent and that the symmetry flow areas between them are more pronounced. Around rotating bodies, the flow is often fairly balanced, with symmetrically extending high-velocity regions outward.

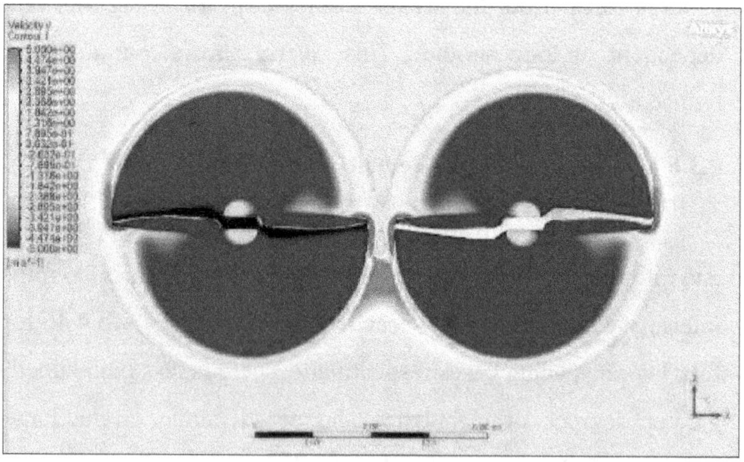

Figure 4.44 LS2.6 Y-Velocity Component at 8000 rpm

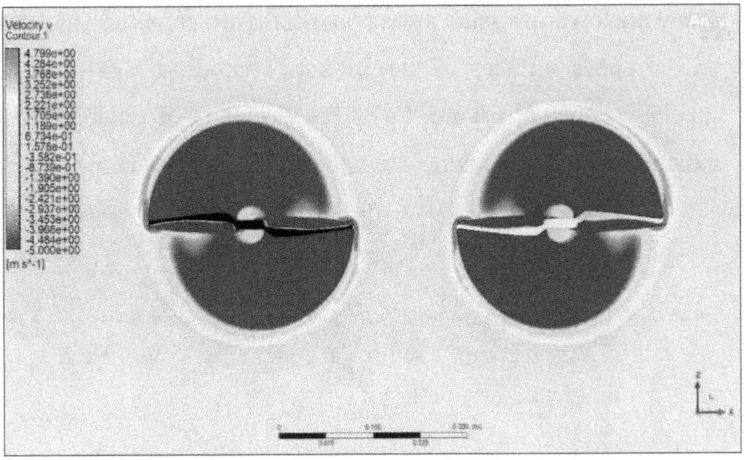

Figure 4.45 LS3.0 Y-Velocity Component at 8000 rpm

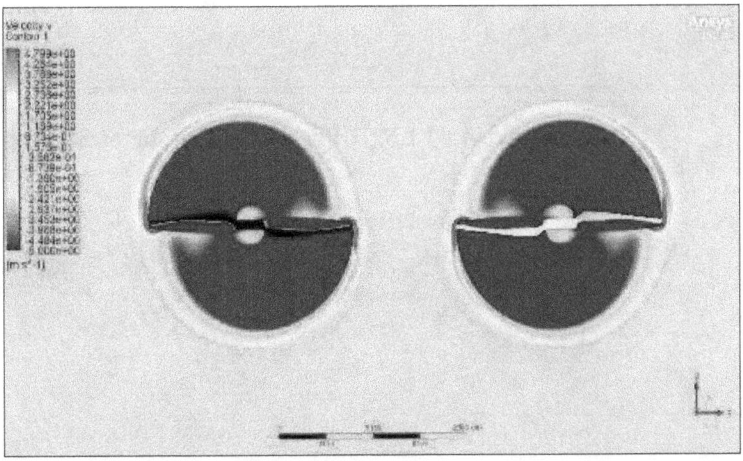

Figure 4.46 LS3.4 Y-Velocity Component at 8000 rpm

As the distance between the rotors is increased, the interaction effects seem mitigated particularly at the lateral spacing 3. After which, there is no further reduced interference effects. The flow pattern almost seems the same. It supported the results obtained at the speed of 5000 rpm.

4.3.7 Lateral Spacing Pressure Contour at 5000 rpm

There are lower-pressure areas beneath the structures and higher-pressure areas on the upper surfaces of the pressure contour at 5000 rpm with 2.2 lateral spacing shown in Figure 4.47. The presence of notable pressure gradients indicates that the structures are under strong aerodynamic forces.

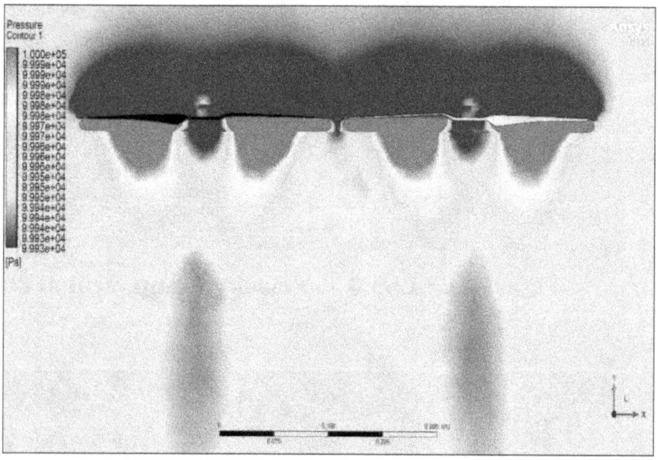

Figure 4.47 LS2.2 Pressure Contour at 5000 rpm

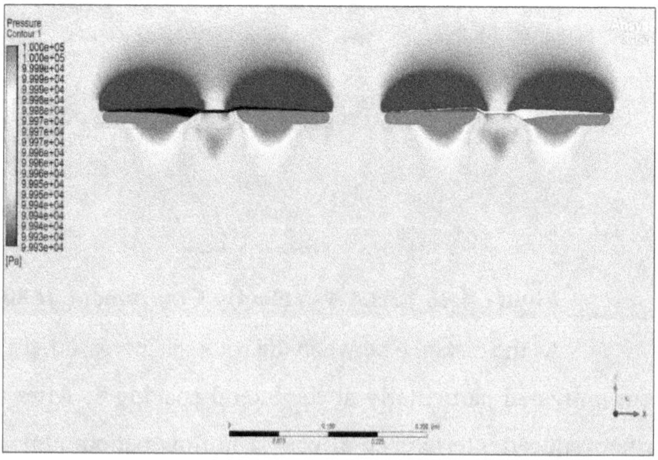

Figure 4.48 LS2.6 Pressure Contour at 5000 rpm

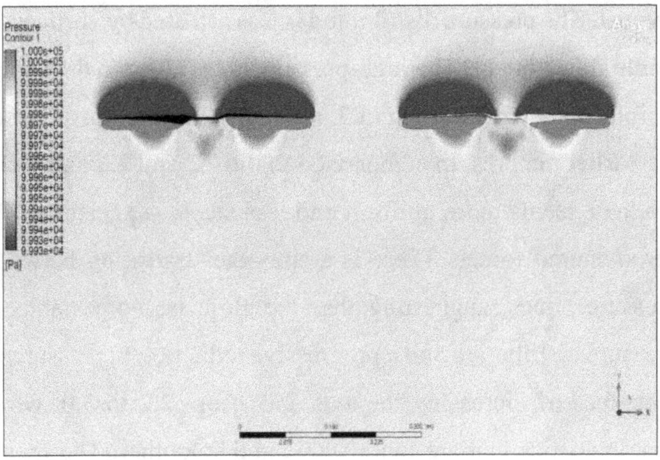

Figure 4.49 LS3.0 Pressure Contour at 5000 rpm

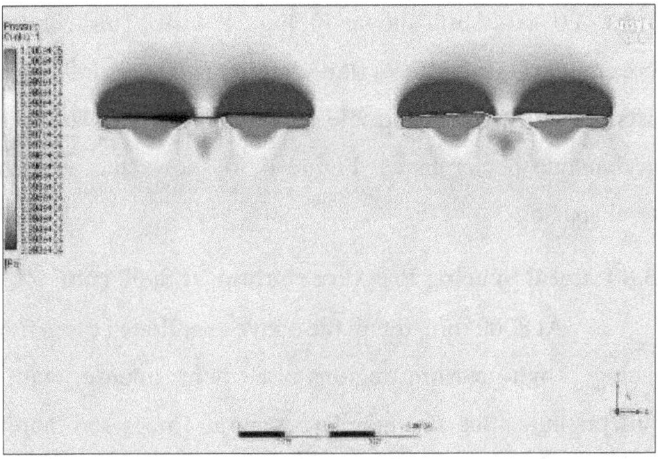

Figure 4.50 LS3.4 Pressure Contour at 5000 rpm

The areas of high pressure correspond with the places where vortices form and influence the flow, resulting in pressure variations. Although they are a little less severe than they were at 2.2 axis shift, high-pressure zones around the upper surfaces are still visible in the pressure contour at 2.6 lateral spacing shown in Figure 4.48. The pressure differential seems more even, indicating reduced aerodynamic forces and more fluid movement, even though it is still present. The pressure distribution is less affected by vortices, suggesting a more stable flow pattern. The high-pressure areas close to the upper surfaces are still visible at a lateral spacing of 3, although they are not as strong as they were in the earlier pictures. In comparison to the 2.2 and 2.6 axis changes, the pressure gradient seems more uniform and less steep, suggesting a further decrease in aerodynamic forces. There is a smoother transition between high- and low-pressure zones, suggesting that the flow is more stable. A more uniform pressure distribution and a progressive reduction in pressure differences are the outcomes of increasing the axis shift from 2.2 to 3.0, which produces flow characteristics that are more stable and smoother. The most stable flow with the least amount of vortex influence on the pressure distribution is displayed by the 3.0 axis shift shown in Figure 4.49. This comparison shows how increasing axis shift affects flow stability and pressure distribution; higher axis shifts lead to more favourable flow characteristics and possibly even better aerodynamic performance. Figure 4.50 shows the pressure contour for the lateral spacing 3.4.

4.3.8 Lateral Spacing Pressure contour at 8000 rpm

At 8000 rpm, for all the above-mentioned cases from 2.2 to 3.4 lateral spacing, High-pressure regions are more intense, with steeper pressure gradients indicating stronger aerodynamic forces and more significant vortex influences and these effects are shown from Figures 4.51 to 4.53. More complex vortex structures will be created. Increasing the RPM from 5000 to 8000 results in more intense high-pressure regions, steeper pressure gradients, and greater aerodynamic forces. The higher RPM leads to stronger vortex influences and more complex flow patterns, indicating more significant pressure differences and enhanced aerodynamic effects.

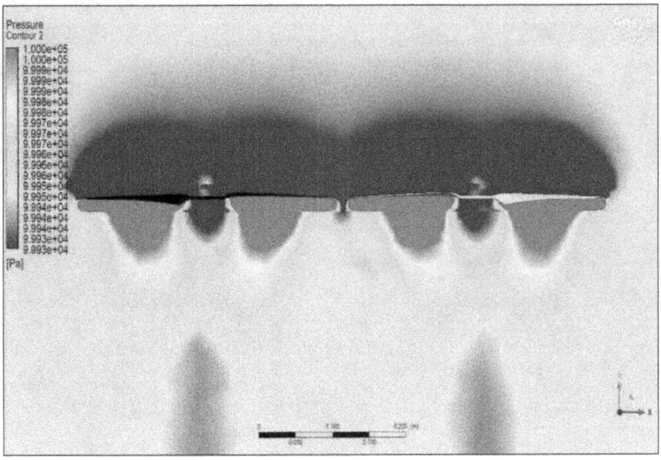

Figure 4.51 LS2.2 Pressure Contour at 8000 rpm

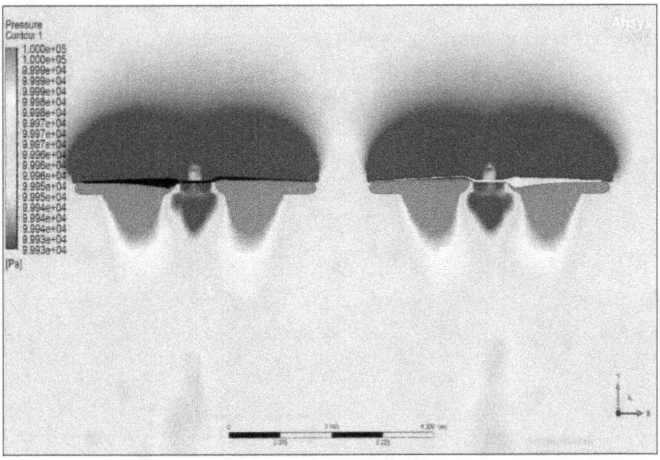

Figure 4.52 LS2.6 Pressure Contour at 8000 rpm

The distribution and strength of the high-pressure areas as well as the steepness of the pressure gradients are the main variations between the cases. A more uniform distribution of aerodynamic forces results from the high-pressure zones spreading out and the pressure gradients becoming less steep as the axis shift rises. This consistency may lead to enhanced functionality and less strain on the structure.

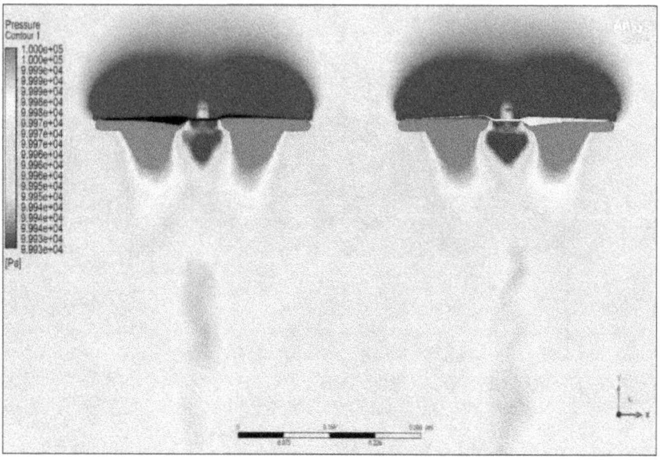

Figure 4.53 LS3.0 Pressure Contour at 8000 rpm

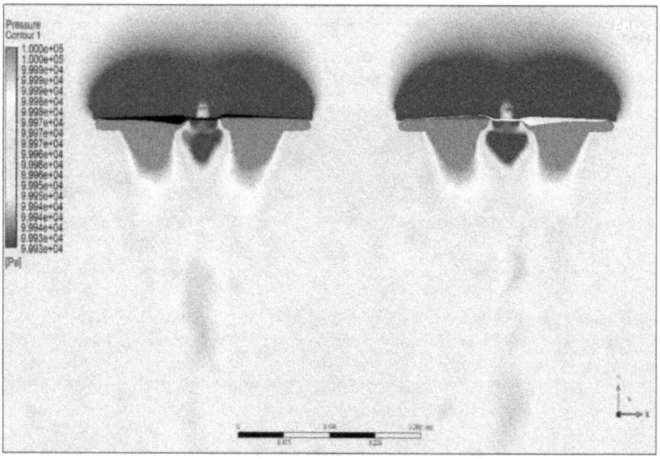

Figure 4.54 LS3.4 Pressure Contour at 8000 rpm

4.3.9 Lateral Spacing Turbulence Kinetic Energy Contours at 5000 rpm

The colour scale in the Figure 4.55 for 2.2 lateral spacing indicates higher levels of TKE when we compare the Q-Criterion at 2.2 and 2.6 lateral spacing. In the 2.2 image, the maximum TKE value is roughly 3.909e+02 m^3/s^2. Lower TKE levels are seen in the image for 2.6, with the highest value reaching approximately 1.422e+02 m^2/s^2. In comparison to the other spacings, the vortex structures in the least 2.2 spacing appear more elongated and stretch farther downstream.

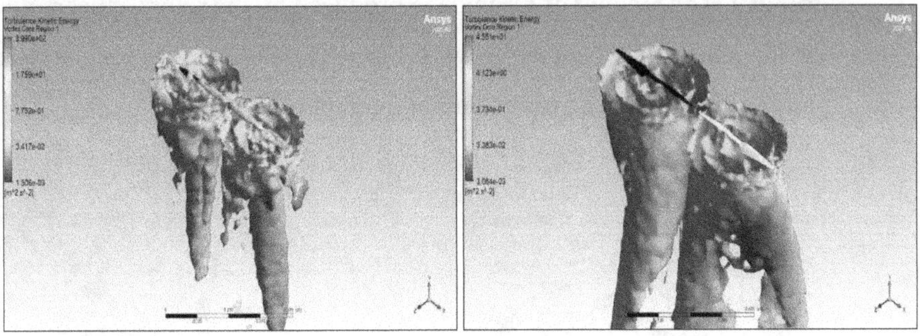

a) Q-Criterion b) Vorticity

Figure 4.55 Lateral Spacing Turbulence Kinetic Energy Contours for LS2.2 at 5000 rpm

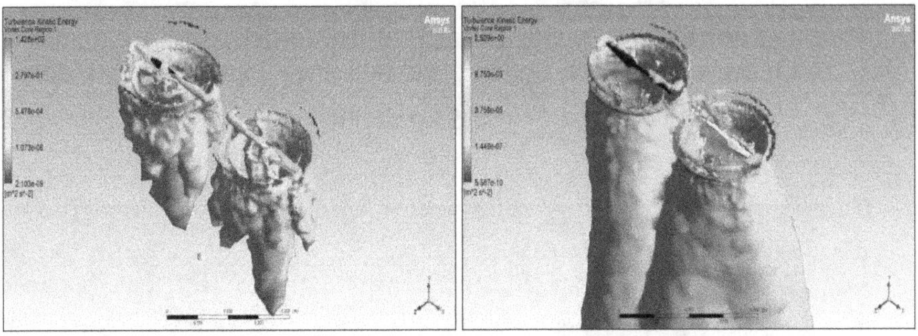

a) Q-Criterion b) Vorticity

Figure 4.56 Lateral Spacing Turbulence Kinetic Energy Contours for LS2.6 at 5000 rpm

This suggests that the turbulence in the LS 2.2 is more intense and noticeable. Weaker turbulence is suggested by the other's shorter and more compact vortex formations. In general, less turbulence means more stable flying conditions. The picture with the 2.6 axis shift distance indicates more stable aerodynamic circumstances, which are good for UAV performance, by displaying lower amounts of turbulence kinetic energy (TKE) shown in Figure

4.56. Reduced turbulent kinetic energy (TKE) translates into less energy lost. Longer flying times and more energy efficiency are possible outcomes for UAVs. Because of its reduced turbulence, the LS 2.6 should be more energy efficient. The amount of turbulence produced by the rotors appears minimal as the lateral spacing distance grows because there is enough space between them, and the vortices of the rotors do not interact.

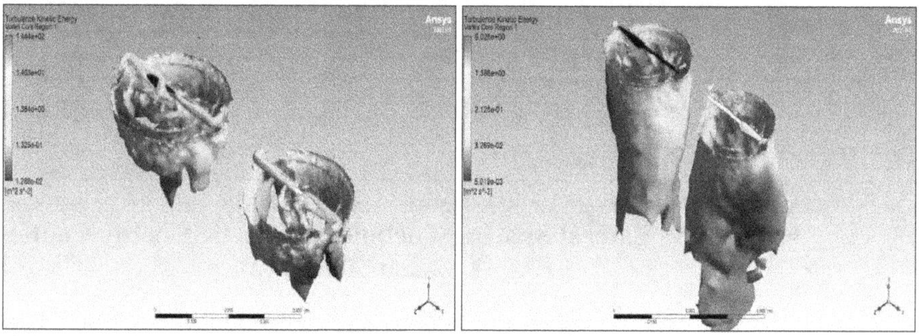

a) Q-Criterion b) Vorticity

Figure 4.57 Lateral Spacing Turbulence Kinetic Energy Contours for LS3.0 at 5000 rpm

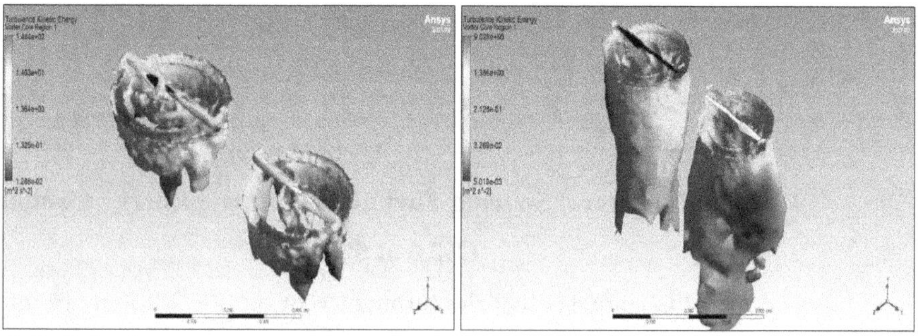

a) Q-Criterion b) Vorticity

Figure 4.58 Lateral Spacing Turbulence Kinetic Energy Contours for LS3.4 at 5000 rpm

While the turbulence in LS 3.4 is less pronounced and more dispersed than in the lower lateral spacing cases, it is still moderate as shown in Figure

4.58. When comparing LS 3.4 and LS 3, both produced less turbulence compared to the remaining cases, with the highest TKE level of approximately 1.444e+02 m^2/s^2. In LS 3, the vortex structures appear compact, and the turbulence is moderate, with primarily green and blue colours indicating lower turbulence. The two examples stated are similar to each other. The TKE levels and general turbulence characteristics are identical for both the 3.0 and 3.4 axis shift lengths. On the other hand, the 3.4 configuration's somewhat more dispersed vortex formations might point to a modestly more stable aerodynamic environment, which is generally advantageous for UAV performance.

This analysis suggests that, because of possibly more stable aerodynamic circumstances, the configuration with an axis shift distance of 3.4 might provide somewhat superior conditions for UAV performance. The changes are negligible, though, and depending on additional particular performance requirements or limitations, either the 3 or 3.4 arrangement may be appropriate. According to the data, there is less turbulence produced by each rotor as the separation between them increases, which results in a more stable functioning. Higher dynamic loads and vibrations are introduced when turbulence is high, which may have an impact on the UAV's structural integrity and longevity. High turbulence produced at 2.2 and 2.6 might also impair the control and manoeuvrability of the UAV, possibly leading to more unpredictable flight behaviour.

The turbulence and vortices could also be comprehended from the vorticity contours; they followed a pattern that was similar to the kinetic energy of turbulence in the Q-Criterion. These contours also demonstrate how the interference effects dominate performance. In 2.2, the high TKE values (up to 4.123e+01) indicate strong turbulence around the upper half of the structures when we compare the vorticity contours for other lateral spacing. The vortex cores are the focal point of the energy. Stronger vortices occur as a result of the more obvious flow separation. This suggests a flow that is more unpredictable and vigorous. Reduced vortices strength with a 2.6 axis shift is shown by decreased maximum TKE values (up to 2.529e+00). With fewer areas of elevated TKE, the flow seems more stable. A change from turbulent to more laminar flow is shown by the decreased turbulence. The vortices appear feeble in axis shifts 3 and 3.4, and the highest TKE values in this image are much lower (up to 1.386e+00) than in axis shifts 2.2 and 2.6. This suggests even less intense turbulence. With weaker rotational motion and less energy concentration surrounding the vortex cores, the vortices are significantly less noticeable. A more stable and laminar flow is indicated by the TKE distribution, which is more uniform and concentrated near the top of the structures. TKE and turbulence intensity are clearly declining as the axis shift increases from 2.2 to 3.0. The flow is extremely low turbulence and highly stable with a 3.0 axis shift. The TKE is uniformly distributed and the vortices are weak, suggesting a nearly laminar flow. This implies that raising the axis shift more will keep the flow stable and reduce turbulence, especially at and after 3. Thus, based on the intended balance between turbulence reduction and system performance, the ideal axis shift can be chosen.

4.3.10 Lateral Spacing Turbulence Kinetic Energy Contours at 8000 rpm

At 2.2 lateral spacing, the Maximum TKE is 3.909e+02, indicating strong turbulence but less than at 8000 RPM. Maximum TKE is 3.997e+02 for 2.2 at 8000 RPM, indicating even stronger turbulence due to the higher rotational speed. With the same axis shift of 2.2, increasing the RPM from 5000 to 8000 leads to much larger turbulence kinetic energy, stronger vortices, and improved flow separation. An increasingly chaotic and dynamic flow pattern results from the higher RPM's induction of increased turbulence and energy dissipation.

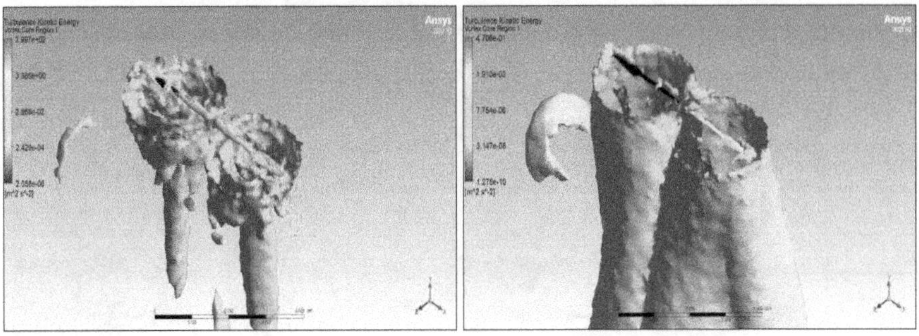

a) Q-Criterion b) Vorticity

Figure 4.59 Lateral Spacing Turbulence Kinetic Energy Contours for LS2.2 at 8000 rpm

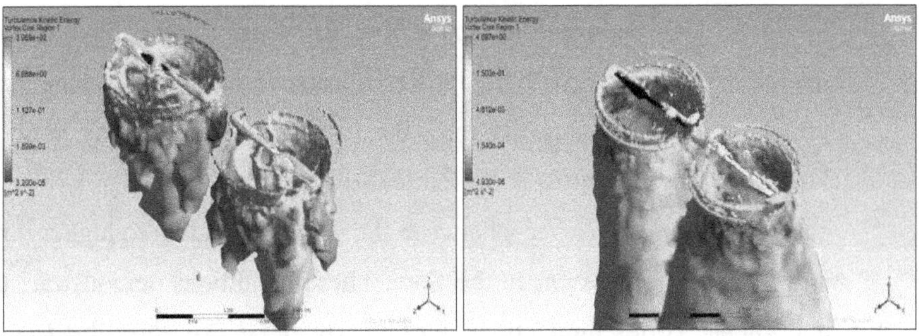

a) Q-Criterion b) Vorticity

Figure 4.60 Lateral Spacing Turbulence Kinetic Energy Contours for LS2.6 at 8000 rpm

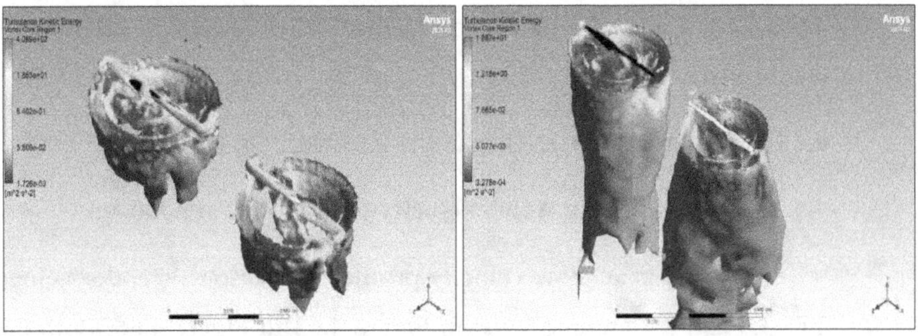

a) Q-Criterion b) Vorticity

Figure 4.61 Lateral Spacing Turbulence Kinetic Energy Contours for LS3.0 at 8000 rpm

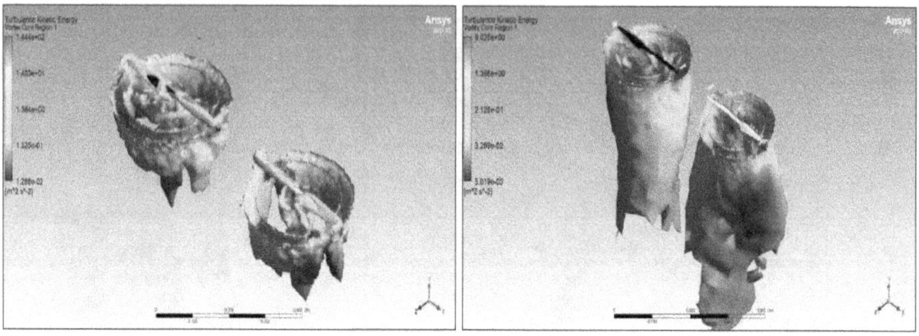

a) Q-Criterion b) Vorticity

Figure 4.62 Lateral Spacing Turbulence Kinetic Energy Contours for LS3.4 at 8000 rpm

This comparison shows how RPM affects the flow features and degree of turbulence, with higher RPM producing more noticeable turbulence effects. With increasing the rpm, the increased turbulence is noticed at all lateral spacing. In vorticity contours also it could be seen clearly that how the strongest interaction effects between the vortices, leading to higher turbulence and instability is present in the flow. These conditions demonstrate complex and intense vortex interactions. Particularly at 2.2 axis shift, due to the closer proximity, more interaction between the rotors is seen. Compared to all other lateral spacing conditions, 3 shows a balance with moderate turbulence and improved stability. The Q-Criterion and vorticity contours are shown in Figures 4.59, 4.60, 4.61 and 4.62 for the cases of LS 2.2, LS 2.6, LS 3.0 and LS 3.4 respectively.

4.4 LATERAL SPACING AXIAL VELOCITY PROFILE

4.4.1 Lateral Spacing Axial Velocity Profile Plot at 5000 rpm

The axial flow velocity profiles for various lateral spacings of 2.2, 2.6, 3.0 and 3.4 are displayed in Figure 4.63 to Figure 4.66. Both the inflow (red lines) and outflow (blue lines) velocity profiles are shown. The outflow velocity for Lateral Spacing 2.2 exhibits two notable peaks that reach 20 m/s.

Between 2 and 5 m/s, the inflow velocity (red line) varies comparatively steadily. Similar pattern with two peaks in the outflow velocity for Lateral Spacing 2.6, which is slightly lower than in the 2.2 is seen.

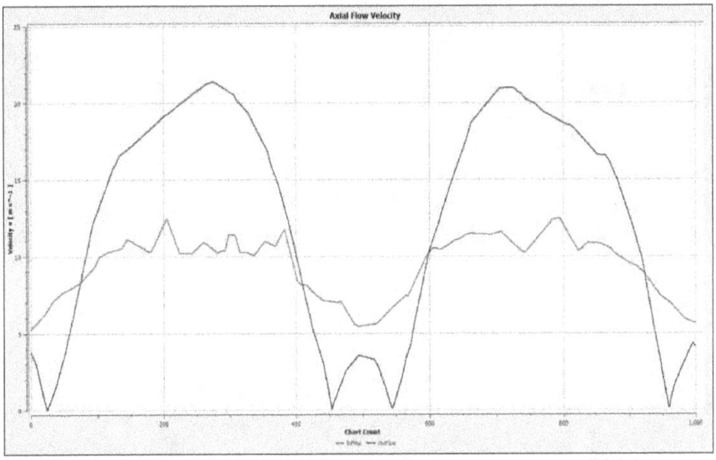

Figure 4.63 LS2.2 Axial Velocity Profile Plot at 5000 rpm

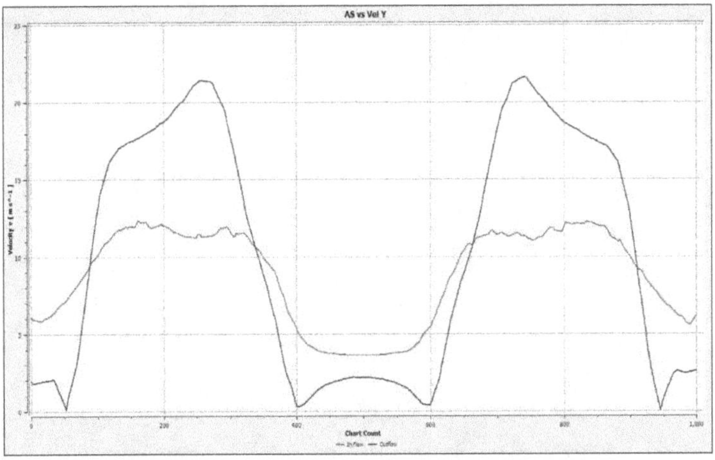

Figure 4.64 LS2.6 Axial Velocity Profile Plot at 5000 rpm

Compared to the previous graph, the inflow velocity exhibits greater stability, with just slight variations between 2 and 5 m/s. The outflow velocity peaks in Lateral Spacing 3 and 3.4 are marginally less than those in the earlier graphs. The input velocity seems more stable but stays within the same range.

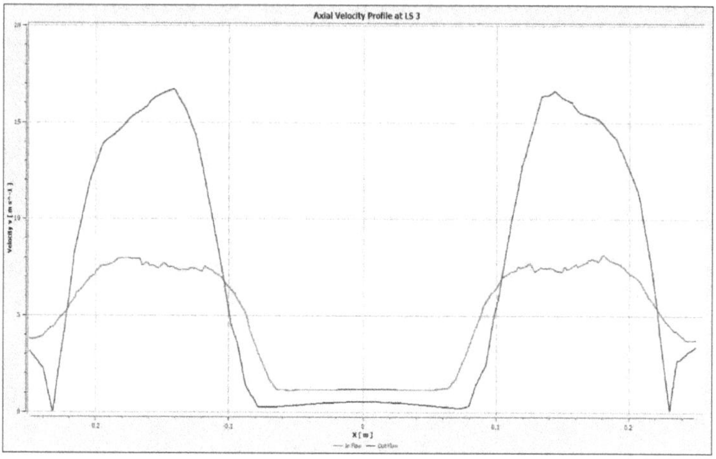

Figure 4.65 LS3.0 Axial Velocity Profile Plot at 5000 rpm

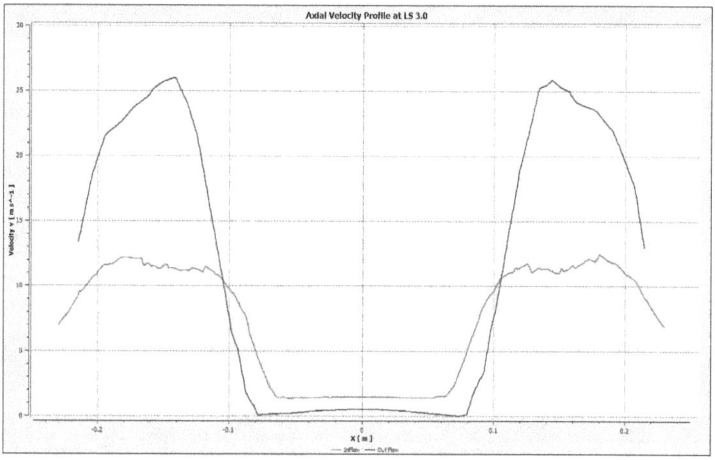

Figure 4.66 LS3.4 Axial Velocity Profile Plot at 5000 rpm

Vortices, or swirling flows, can have a substantial impact on the velocity profiles found in fluid flows when analysing the axial flow velocity profiles in the context of vortices interaction. Usually, variations in the fluid's pressure and velocity cause vortices to form. Figures demonstrate that the outflow has high peaks, which may be a sign that a vortex is forming. The places where the fluid is accelerating, frequently as a result of vortices, are indicated by the velocity peaks. When vortices are periodically shed off an object's surface, this phenomenon is known as vortex shedding. The outflow peaks' periodic nature in the figure's points to the possibility of vortex shedding. The axial velocity may be affected by the interplay of vortices shed from various locations. Stronger vortex interactions are anticipated to produce high velocity peaks in the outflow, and closer lateral spacing may produce more pronounced vortex interactions, which would explain the higher velocity peaks seen in the 2.2 spacing. Closer vortex interaction results in a higher concentration of energy.

The velocity profile may be impacted by vortices that decay and lose energy as they interact and move downstream. Given that the energy of interacting vortices disperses across a broader area at larger lateral spacings (2.6 and 3), this could account for the slightly lower outflow peaks. Vortices produce a smoother velocity profile because they have greater area to release energy. The comparatively constant input velocity suggests that vortex interactions mostly affect the outflow. Less disturbance to the inflow velocity indicates that the inflow zone is upstream of important places where vortex formation and interaction occur. The interaction and vortex generation are mostly influenced by the distance between the rotors. A similar pattern is also observed at lateral spacing 3.4. These interactions grow less powerful with increasing lateral spacing, which lowers the outflow's peak velocities. It is clear from the comparatively stable inflow that the outflow region is the principal target of the vortex interactions.

4.4.2 Lateral Spacing Axial Velocity Profile Plot at 8000 rpm

Higher outflow velocity peaks are observed at 5000 RPM due to stronger and more coherent vortices than in the axial velocity profiles at 8000 RPM. Lower outflow velocity peaks are the result of more turbulent and weaker individual vortices caused by the increased shedding frequency at 8000 RPM shown Figure 4.67 to 4.70 Peak outflow velocities may be decreased as a result of more turbulent and less coherent flow patterns brought on by the faster rotating speed, which also probably causes more rapid vortex shedding. Strong vortex generation is indicated by the outflow, which peaks at 2.2 lateral spacing at about 10 m/s. Areas of significant vortex interaction are suggested by the peaks.

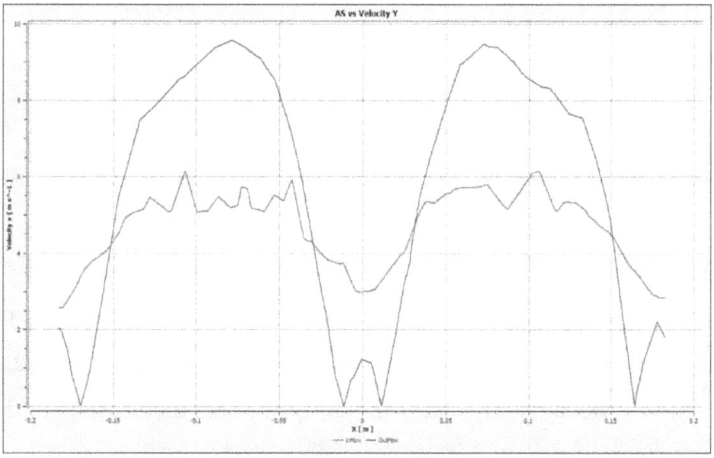

Figure 4.67 LS2.2 Axial Velocity Profile Plot at 8000 rpm

The inflow fluctuates between 2 m/s and 5 m/s, indicating a reasonably consistent inflow. Less influence from vortex generation in the inflow zone is suggested by the stability. At 2.6, a pattern that is similar to the previous graph for 2.2 but with somewhat smaller peaks can be seen. It suggests slightly weaker vortex interactions than at spacing of 2.2. The inflow is constant between 2 and 5 m/s. A steady inflow velocity indicates stability upstream.

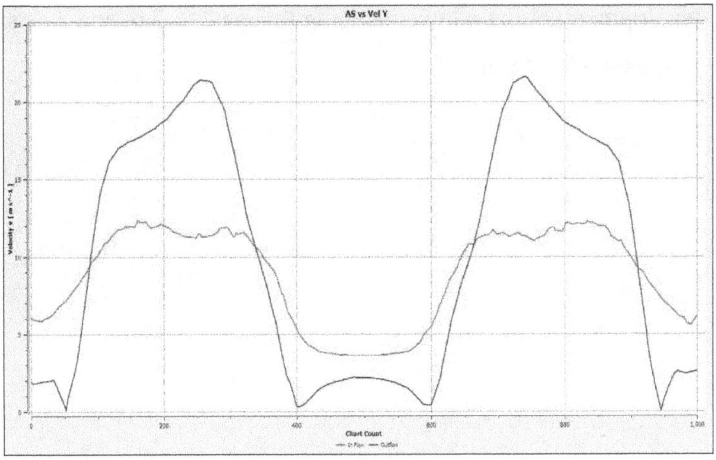

Figure 4.68 LS2.6 Axial Velocity Profile Plot at 8000 rpm

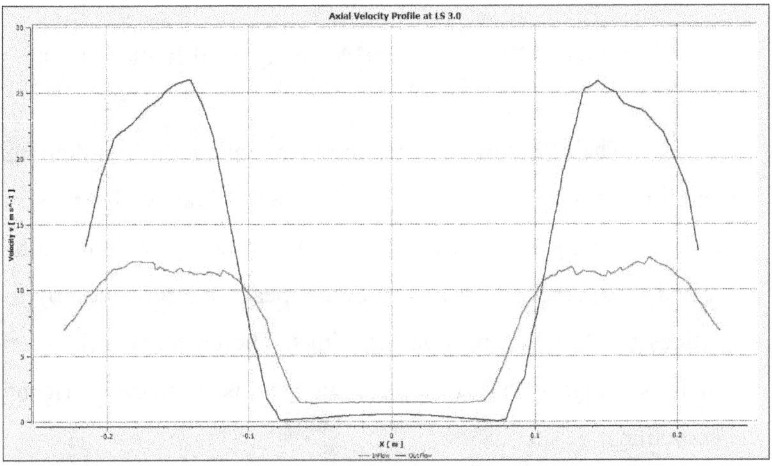

Figure 4.69 LS3.0 Axial Velocity Profile Plot at 8000 rpm

In lateral spacing 3, the outflow peaks at roughly 15 m/s, which denotes a moderately strong vortex. Variations in the kinetics of vortex interaction are suggested by slightly higher peaks than 2.6 spacing. Outflow peaks at 3.4 in Figure 4.70, approximately 15 m/s, indicating a strong vortex development, but at a lower spacing than at 2.2, 2.6, and 3 as shown in Figures 4.67, 4.68 and 4.69.

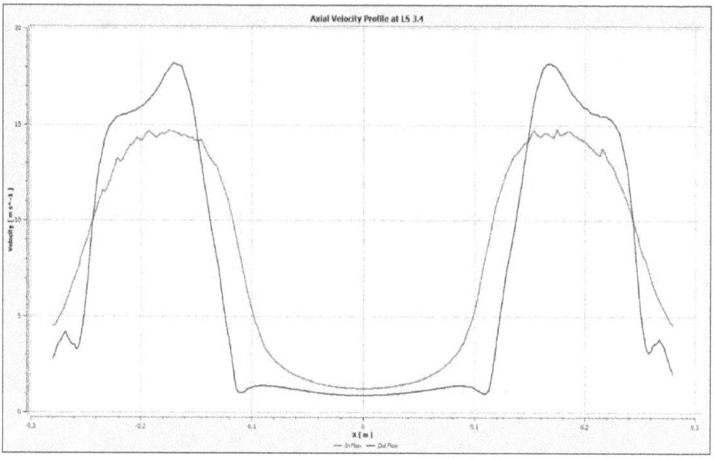

Figure 4.70 LS3.4 Axial Velocity Profile Plot at 8000 rpm

The effect of spacing on vortex interactions is demonstrated by the axial flow velocity profiles at 8000 RPM across various lateral spacings. Higher outflow velocity peaks are the outcome of stronger, more coherent vortices created by smaller spacings. Moderate peaks are produced by more scattered vortices resulting from larger spacings. The constant inflow velocities at all spacings suggest that the outflow area is primarily affected by vortex interactions.

4.5 AXIAL SPACING CONTOURS AND PLOTS

4.5.1 Axial Spacing Velocity Contours for 5000 rpm

Figures 4.71, 4.72 and 4.73 show the velocity contours of H0.5-V0.5, H1-V1 and H1.5-V1.5 at 5000 rpm. In the mentioned first case, the regions around the rotors show high velocity magnitudes (indicated by red and yellow colours). Due to the close axial and downstream spacing, there is a strong interaction between the vortices generated by each rotor. This close proximity causes the vortices to merge quickly, leading to complex and turbulent flow patterns. The wake regions behind each rotor show significant drops in velocity, indicated by green and blue colours.

a) Downstream

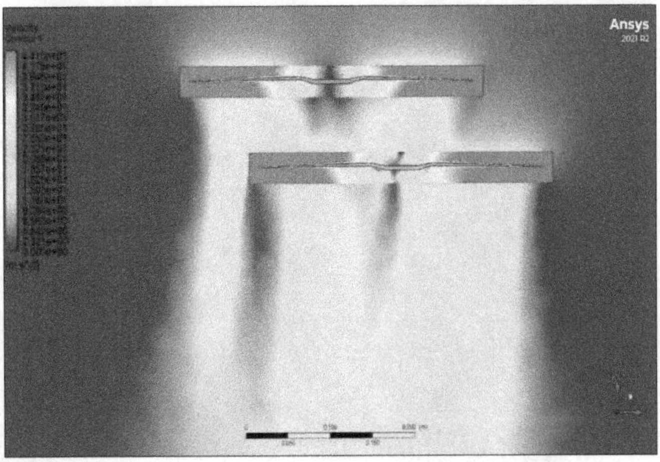

b) Close-up View

Figure 4.71 H0.5-V0.5 Velocity Contour at 5000 rpm

The wake recovery is limited because of the close spacing. The wakes do not have sufficient distance to dissipate and recover before interacting with the next rotor. This results in high turbulence and less efficient flow downstream of each rotor.

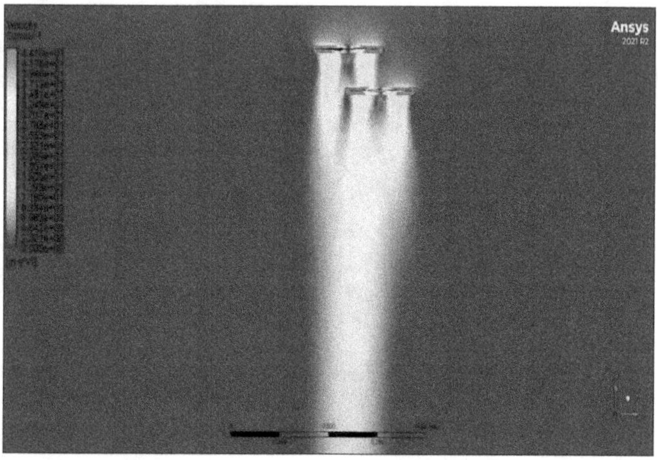

a) Downstream

b)

c) Close-up View

Figure 4.72 H1-V1 Velocity Contour at 5000 rpm

In the second case, the high-velocity regions around the rotors are still prominent, but the wakes show a better distribution and less severe drops in velocity compared to the first image. With increased spacing, the interaction between the vortices is reduced. The vortices still interact but have more space to develop before merging, leading to more organized flow patterns.

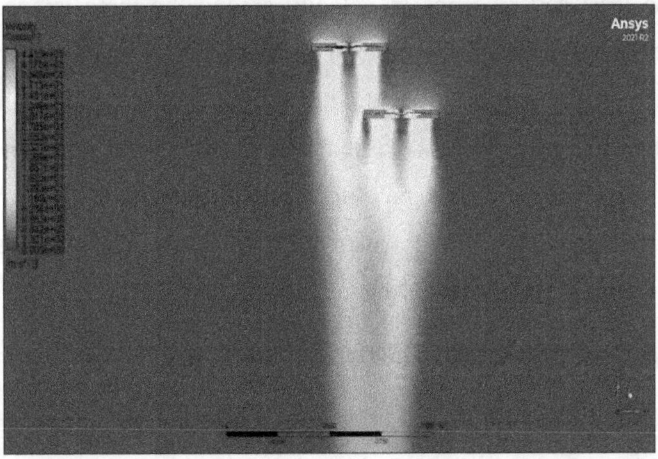

a) Downstream

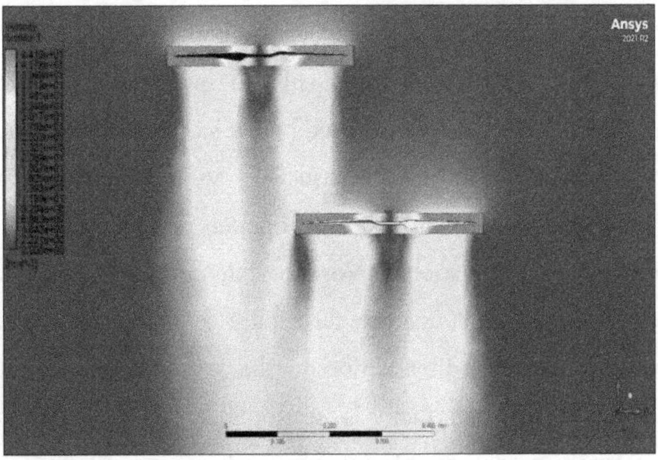

b) Close-up View

Figure 4.73 H1.5-V1.5 Velocity Contour at 5000 rpm

The wake recovery is improved compared to the first image. The wakes have more distance to dissipate and recover, reducing turbulence and improving flow efficiency downstream. There is a clearer separation between the wakes of each rotor, indicating better performance. At H1.5-V1.5, The high-velocity regions are well-defined around the rotors, and the wakes show the least severe drops in velocity among the three images. The interaction between the vortices is minimal due to the larger spacing. The vortices are more distinct and organized, leading to smoother flow patterns and less turbulence. The wake recovery is the best among the three configurations. The wakes have ample distance to dissipate and recover, leading to minimal interaction and turbulence. The flow downstream of each rotor is more stable and efficient, indicating optimal performance.

4.5.2 Axial Spacing Velocity Contours for 8000 rpm

Figures shown below show how the higher speed affects the rotor's performance noticeably when the velocity contours are compared to its rotational speed. As may be predicted given the greater RPM, the high-velocity areas surrounding the rotor blades are noticeable. The 8000 rpm H0.5-V0.5, H1-V1, and H1.5-V1.5 are displayed in Figures 4.74 through 4.76. The interference effects are explained below based on the distance. Strong vortices are produced in the first scenario H0.5-V0.5 as a result of the close spacing. Where vortices form around the propellers is indicated by high-velocity zones (red and yellow regions). Vortices interact significantly with one another. Because wakes can't disperse far enough to interact downstream, limited wake recovery happens. In addition to the vortices' reduced strength, moderate interaction between them is seen under H1-V1 conditions. Partial wake recovery is made possible by increased spacing, which lowers turbulence. A greater balance between interaction and wake recovery is offered by the 1R spacing. Enhanced wake recovery as compared to the initial configuration, which improved performance and decreased turbulence. Due to their greater separation, vortices are more definite and less strong. There is a noticeable distinction between the propeller wakes in high-velocity areas.

a) Downstream

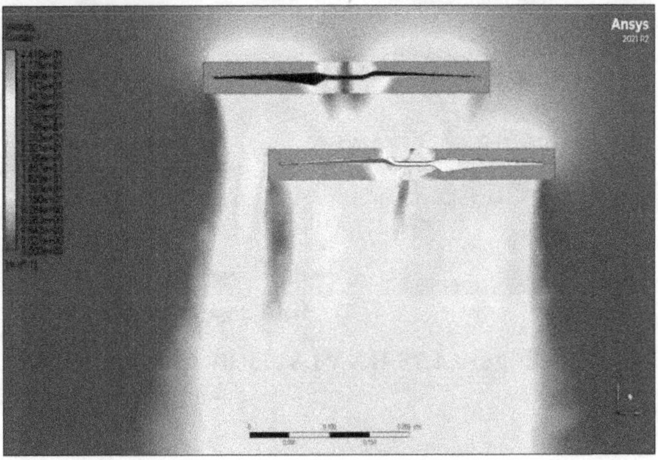

b) Close-up View

Figure 4.74 H0.5-V0.5 Velocity Contour at 8000 rpm

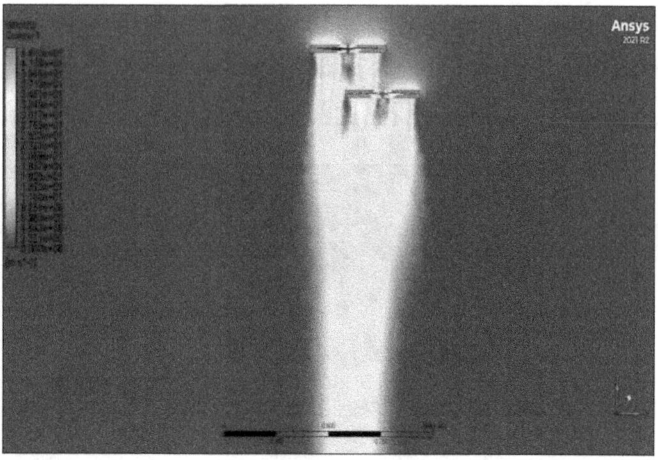

a) Downstream

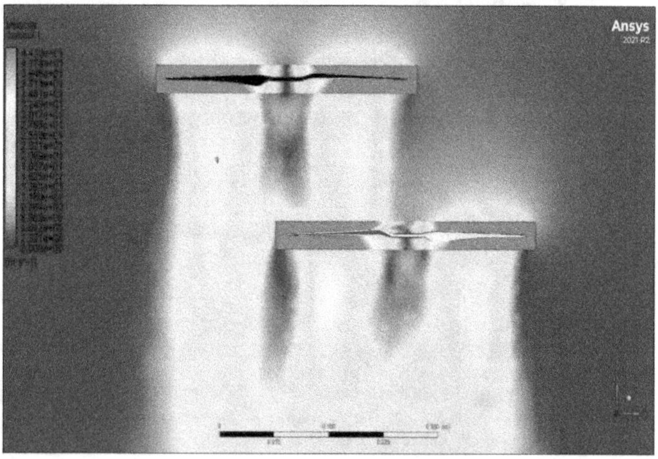

b) Close-up View

Figure 4.75 H1-V1 Velocity Contour at 8000 rpm

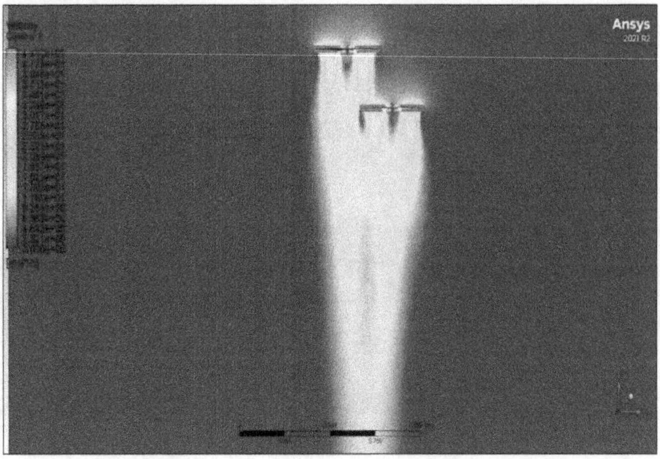

a) Downstream

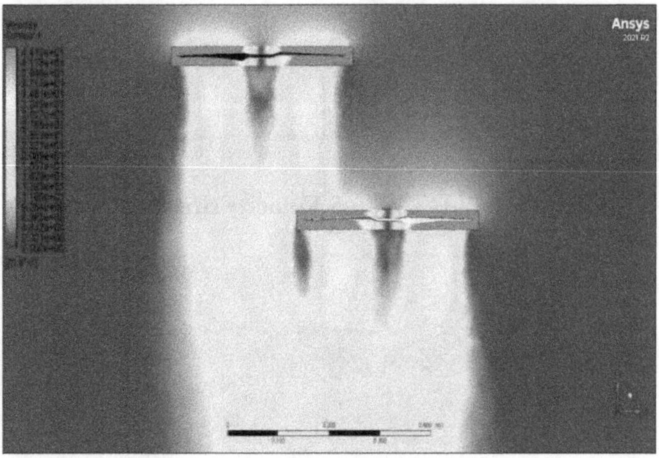

a) Close up View

Figure 4.76 H1.5-V1.5 Velocity contour at 8000 rpm

Due to the wide spacing, there is less turbulence and more stability in the flow field, with minimal contact between the vortices. Less turbulence, greater wake recovery, and enhanced performance are the outcomes. The similar results are only seen at 8000 rpm but with greater intensity.

4.5.3 Axial Spacing Velocity Streamline for 5000 rpm

The streamline patterns for the previously mentioned combinations are shown in figures 4.77 to 4.79, correspondingly, as H0.5-V0.5, H1-V1, and H1.5-V1.5.

Figure 4.77 H0.5-V0.5 Velocity Streamline plot at 5000 rpm

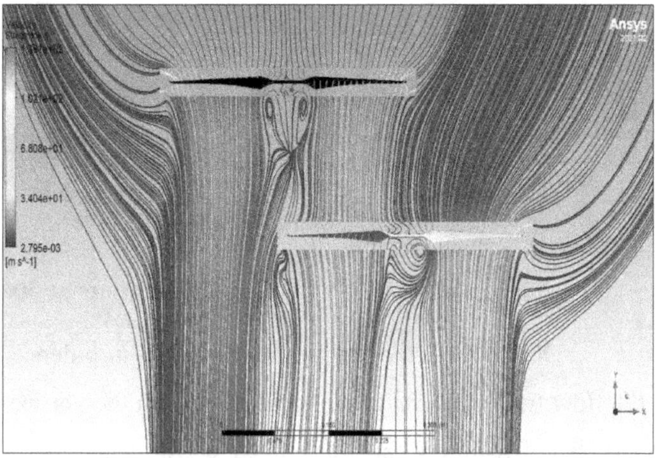

Figure 4.78 H1-V1 Velocity Streamline plot at 5000 rpm

Figure 4.78 shows a substantial vortex generation due to the streamlines being closely packed around the propellers. Tight whirling patterns are indicative of strong vortex interaction and considerable strength. Because of their close spacing, the two propellers' streamlines interact significantly. A very turbulent area is produced when the trailing propeller is immediately impacted by the wake from the leading propeller.

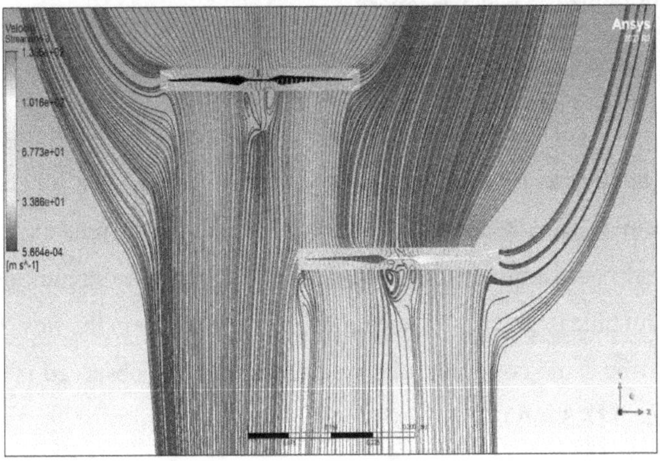

Figure 4.79 H1.5-V1.5 Velocity Streamline plot at 5000 rpm

Their lack of dissipation distance causes them to come into direct contact with the propeller's flow at the trailing end. Compared to the first image, Figure 4.78's streamlines are more dispersed, suggesting a moderate vortex strength. Because of the greater spacing, the vortices are still there but are less intense. The two propellers' streamlines interact moderately with one another. When the wakes contact with the trailing propeller's flow, they have time to grow and partially recover, which lessens turbulence. It offers a more harmonious combination of wake recovery and interactivity. improved wake recovery in comparison to the initial configuration, which results in a flow field that is more stable and has less turbulence. The streamlines in Figure 4.79 are smoother and more widely spaced, suggesting well-defined but somewhat strong vortices. Minimal interaction is evident in the streamlines, indicating minimal interaction between vortices. Because of their great separation, there is very little interaction between the streamlines of the two propellers. There is less turbulence and a more stable flow field with discrete wake areas. This arrangement improves overall aerodynamic performance, reduces turbulence, and improves wake recovery.

4.5.4 Axial Spacing Velocity Streamline for 8000 rpm

The relationship between the streamlines with respect to the spacing at 8000 rpm are shown in figures 4.80, 4.81 and 4.82 for the mentioned three cases in the same order. For H0.5-V0.5 at 8000 rpm, the maximum velocity observed is around 1.604e+02 m/s (160.4 m/s). The streamlines show a gradient from blue to green, suggesting moderate velocity in the flow field but 5000 rpm for the same condition, the maximum velocity observed is around 1.383e+02 m/s (138.3 m/s).

Figure 4.80 H0.5-V0.5 Velocity Streamline plot at 8000 rpm

Figure 4.81 H1-V1 Velocity Streamline plot at 8000 rpm

The streamlines exhibit a gradient from blue to yellow and red near the propeller blades, indicating higher velocities. The UAV rotors at 8000 RPM generate stronger vortices with higher velocity magnitudes, leading to more intense aerodynamic forces and higher turbulence.

Figure 4.82 H1.5-V1.5 Velocity Streamline plot at 8000 rpm

Vortices are created immediately behind the leading propeller blades where the streamlines start to swirl and separate. The trailing propeller also generates vortices immediately behind its blades, but these are heavily influenced by the wake from the leading propeller. Significant vortex interaction occurs due to the close spacing, causing complex swirling patterns almost immediately behind both sets of propeller blades. The maximum velocity range obtained for the cases are 160.4m/s, 157.7m/s and 157.2m/s. However, the space is increased from 1R to 1.5R, the variation seems minimal in the maximum velocity. As the distance is increased in both the axial and lateral spacing, the maximum velocity range is reduced and weaker vortices could also be seen clearly from the figure.

4.5.5 Axial Spacing Pressure Contours for 5000 rpm

The pressure distribution around the propellers at 5000 rpm for H0.5-V0.5, H1-V1, and H1.5-V1.5 is depicted in Figures 4.83, 4.84, and 4.85. The rotors in the first example are arranged both laterally and axially closely. Red areas of high pressure are observed close to the rotor tips. Regions of low pressure (blue) are visible in the wake behind the rotors. There is a great deal of interaction between the rotors, which results in complex pressure gradients. The pressure varies between 9.993e+04 Pa and 1.000e+05 Pa for three examples. Intense contact results in increased aerodynamic drag and turbulence. Though less effective, suitable for great mobility. Rotors are arranged considerably apart both laterally and axially at H1-V1. high-pressure areas close to the tips of the rotors, as in the earlier arrangement. Wake zones at low pressure are better structured and less affected by rotor interaction. As opposed to the closer spacing, there is less interaction, which leads to more consistent pressure patterns. When compared to closer spacing, balanced performance with increased efficiency can be attained. The rotors are farther apart laterally and axially at H1.5-V1.5.

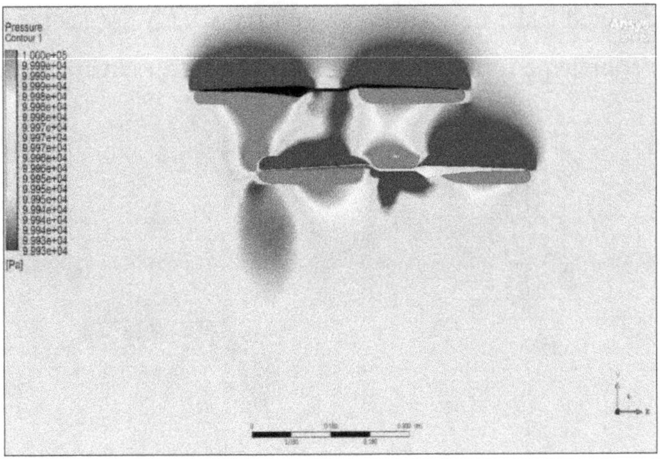

Figure 4.83 H0.5-V0.5 Pressure contour at 5000 rpm

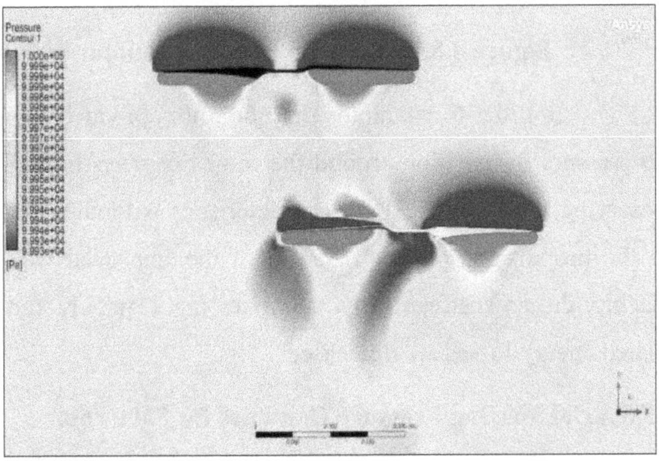

Figure 4.84 H1-V1 Pressure contour at 5000 rpm

Low-pressure areas show less interaction between the rotors because they are more pronounced and isolated. Greater spacing permits the flow to stabilize, resulting in distinct pressure patterns with less disturbance. Achieving the best possible aerodynamic performance with the least amount of drag and turbulence is appropriate for steady and effective flight operations. The more structured wake flow at H1-V1 and H1.5-V1.5 lessens the instability that can deteriorate performance, especially in turbulent circumstances.

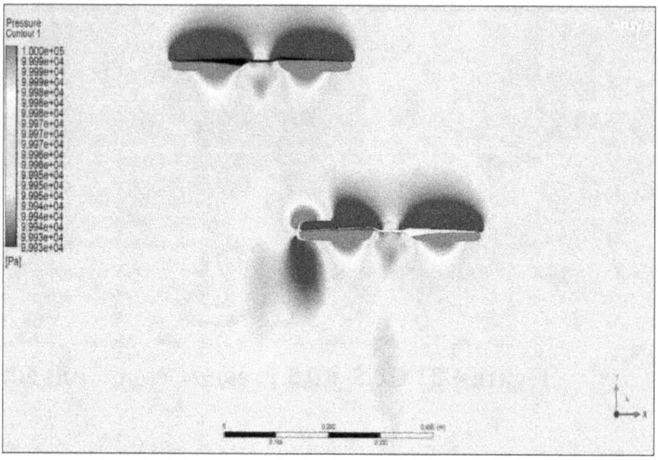

Figure 4.85 H1.5-V1.5 Pressure contour at 5000 rpm

In order to guarantee that the rotors produce enough thrust and lift, the pressure distribution around the rotor tips stays high. This is essential to preserving the intended flight characteristics without putting too much strain on the propulsion system. Because of the improved wake recovery at this spacing, the downstream flow stabilizes more quickly and wake turbulence-related energy losses are decreased.

4.5.6 Axial Spacing Pressure Contours for 8000 rpm

It offers greater thrust and lift at 8000 rpm compared to 5000 rpm at all spacing situations, but at the drawback of more turbulence and aerodynamic inefficiencies, especially for the downstream rotor. Applications requiring strong thrust may find this arrangement appropriate, but careful control of the aerodynamic effects is necessary. The upstream rotor functions in undisturbed air in all configurations, preserving steady aerodynamic loads and effective operation. The undisturbed flow conditions lead to high efficiency and optimal thrust generation. The overall thrust efficiency varies because of the downstream rotor, which operates under the influence of the upstream rotor. Due to near wake interaction, it experiences considerable turbulence and lower pressure regions at H0.5-V0.5 as shown in Figure 4.86.

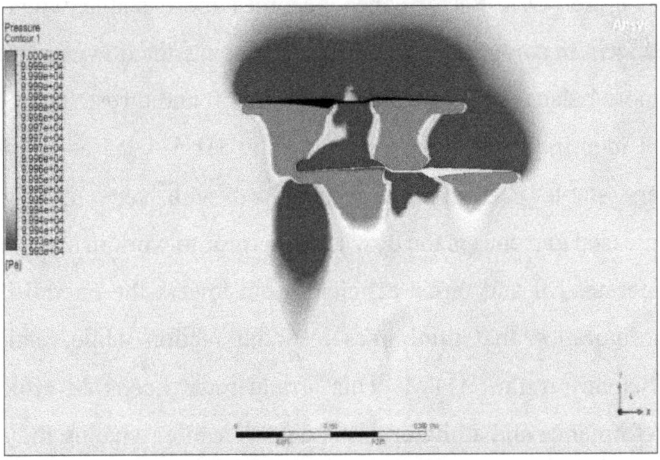

Figure 4.86 H0.5-V0.5 Pressure contour at 8000 rpm

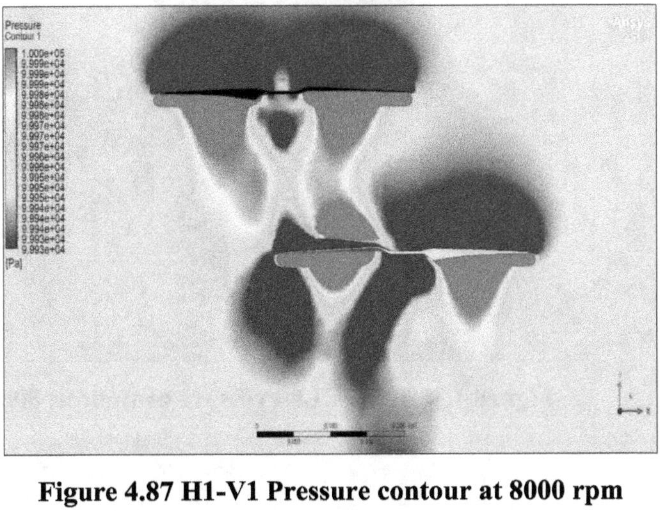

Figure 4.87 H1-V1 Pressure contour at 8000 rpm

Reduced aerodynamic efficiency and higher turbulence are the results of strong wake effects. There are more noticeable losses in lift and thrust when the downstream rotor runs in a highly disturbed flow. Experiences moderate turbulence and better-organized low-pressure areas at H1–V1 shown in Figure 4.87. Performance becomes more stable when wake effects are reduced. In comparison to the closer spacing, the downstream rotor experiences a more balanced flow, which improves lift and thrust. There is little turbulence and identifiable low-pressure zones at H1.5–V1.5 as shown in Figure 4.88. Very stable performance is achieved with very little wake effects. The increased gap causes the downstream rotor to work in a very stable flow, which generates lift and thrust efficiently but lowers the overall thrust density. The configuration that minimizes rotor interaction while retaining aerodynamic efficiency is the H1-V1. This arrangement keeps an effective aerodynamic performance and a higher thrust density while reducing drag and turbulence.

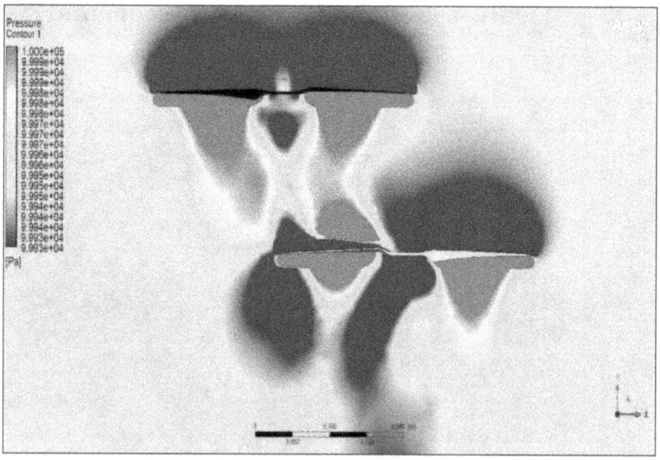

Figure 4.88 H1.5-V1.5 Pressure contour at 8000 rpm

4.5.7 Axial Spacing Turbulence Kinetic Energy Contours at 5000 rpm

At 5000 rpm, Lower TKE values indicate reduced turbulence as shown in Figures 4.89, 4.90, and 4.91 for the cases of H0.5-V0.5, H1-V1 and H1.5-V1.5.

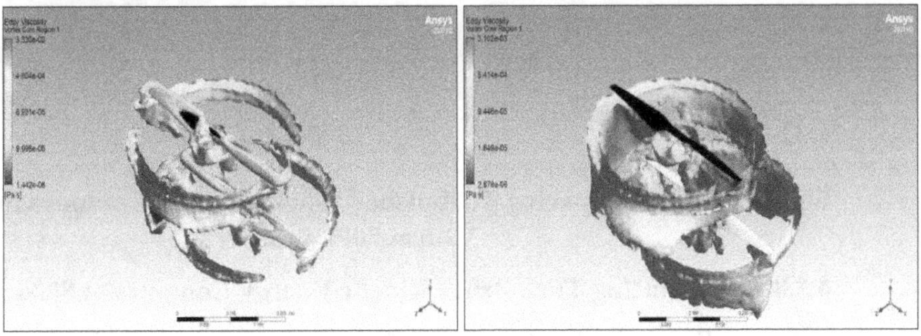

a) Q-Criterion b) Vorticity

Figure 4.89 Axial Spacing Turbulence Kinetic Energy Contours for H0.5-V0.5 at 5000 rpm

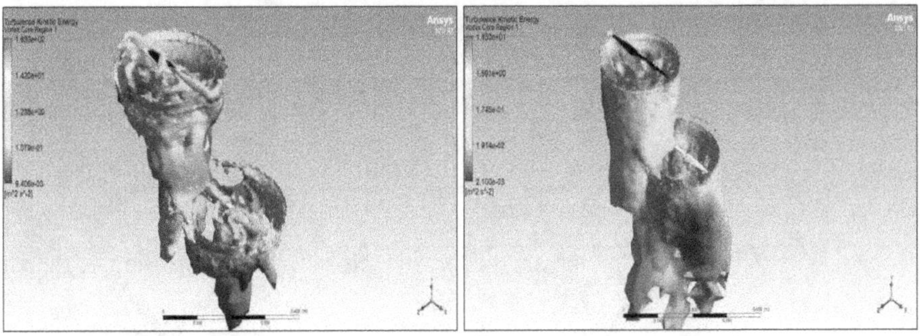

a) Q-Criterion b) Vorticity

Figure 4.90 Axial Spacing Turbulence Kinetic Energy Contours for H1-V1 at 5000 rpm

The rotor generates weaker vortices, leading to less wake turbulence compared to 8000 rpm. However, as the axial and lateral spacing between the rotor increases, the vortex interaction is reduced.

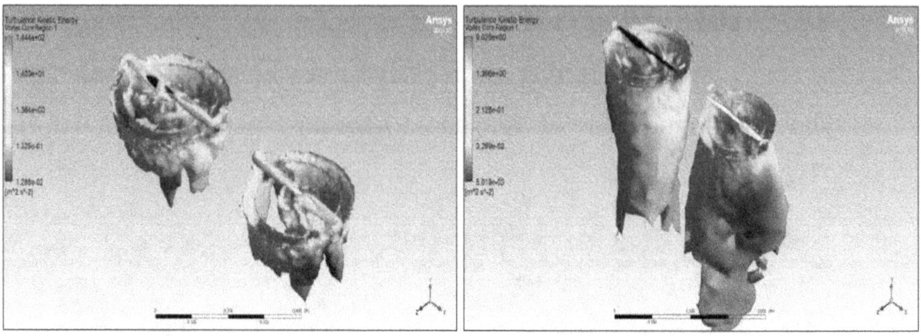

a) Q-Criterion b) Vorticity

Figure 4.91 Axial Spacing Turbulence Kinetic Energy Contours for H1.5-V1.5 at 5000 rpm

4.5.8 Axial Spacing Turbulence Kinetic Energy Contours at 8000 rpm

From Figure 4.92 to 4.94, the Q-criterion and vorticity contours of H0.5-V0.5, H1-V1 and H1.5-V1.5 at 8000 rpm are shown. This image shows the eddy viscosity distribution in the vortex core region for UAV rotors with an axial spacing of 0.5. Eddy viscosity is an indicator of the turbulent mixing and momentum diffusion in the flow.

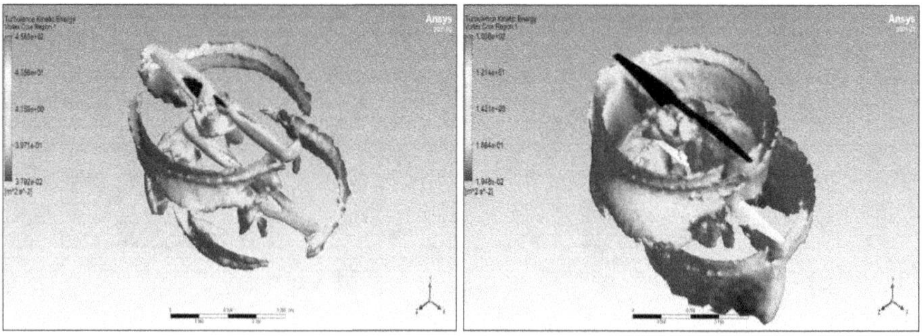

a) Q-Criterion b) Vorticity

Figure 4.92 Axial Spacing Turbulence Kinetic Energy Contours for H0.5-V0.5 at 8000 rpm

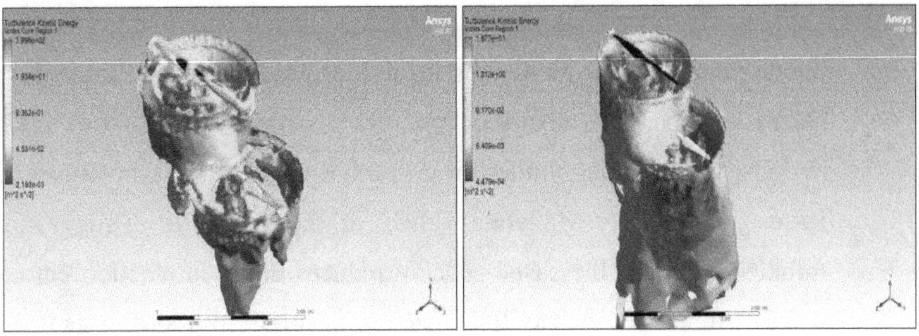

a) Q-Criterion b) Vorticity

Figure 4.93 Axial Spacing Turbulence Kinetic Energy Contours for H1-V1 at 8000 rpm

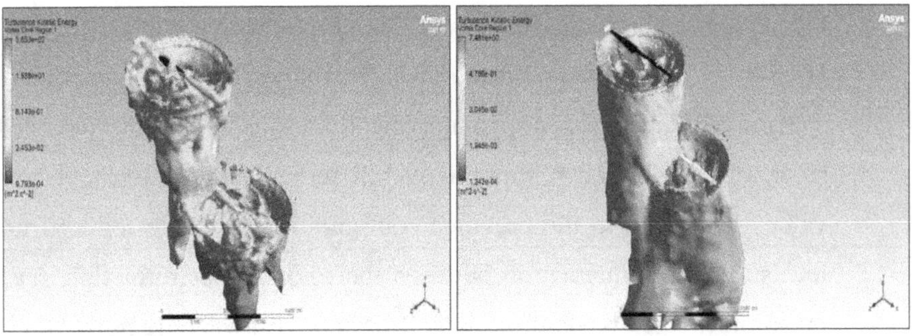

a) Q-Criterion b) Vorticity

Figure 4.94 Axial Spacing Turbulence Kinetic Energy Contours for H1.5-V1.5 at 8000 rpm

The simulation depicts how closely spaced rotors influence each other's wake and turbulence characteristics. At H0.5-H0.5, higher values of eddy viscosity near the rotor tips indicate stronger turbulent mixing. Rotors are placed very close to each other. The close proximity of the rotors results in significant interaction between their vortices, leading to complex flow structures. The downstream rotor operates in the disturbed flow created by the upstream rotor. In upstream rotor, the velocity near the rotor tips reaches high values due to the strong induced flow and rotor blade movement.

In downstream Rotor, the velocity magnitude is highly influenced by the turbulent wake from the upstream rotor, leading to significant fluctuations. Near rotor tips, the velocity magnitudes can reach up to 80-100 m/s. In wake regions, velocity magnitudes drop to 20-40 m/s due to turbulence and energy losses. At H1-V1, Moderate values in the vortex core suggest balanced turbulence levels. Increased spacing reduces direct interaction between rotor wakes compared to the 0.5 spacing, leading to more organized vortex structures. Moderate interaction between the rotors. The downstream rotor experiences a partially recovered flow from the upstream rotor's wake. Near rotor tips, the velocity magnitudes can reach up to 80-100 m/s. In wake regions, the velocity magnitudes are around 30-50 m/s, with moderate fluctuations. At H1.5-V1.5, the image shows the TKE distribution for UAV rotors with an axial spacing of 1.5. Increased spacing typically results in less interference between rotor wakes, which can improve overall aerodynamic efficiency. Lower TKE values compared to the previous configurations indicate reduced turbulence levels. Minimal interaction between the rotors. Near rotor tips, the velocity magnitudes can reach up to 80-100 m/s. In wake regions, the velocity magnitudes are around 40-60 m/s, with minimal fluctuations. The downstream rotor operates in a more recovered and stabilized flow. The larger spacing allows the vortices to develop more independently, minimizing the interaction and leading to more stable flow patterns.

In vorticity figures, the same pattern is shown at the above mentioned three cases, at H0.5-V0.5, The proximity results in complex and overlapping vortex structures. TKE values range from approximately 0 to 1.068e+02 m^2/s^2, with high values near the rotor tips and in the immediate wake. At H1-V1, increased spacing allows some recovery of the flow, leading to reduced interaction between rotor wakes. TKE values range from approximately 4.47e-04 to 1.877e+01 m^2/s^2, with moderate values in the wake regions. At H1.5-

V1.5, The larger spacing allows for substantial recovery of the flow, resulting in a stable and organized velocity field. TKE values range from approximately 1.243e-04 to 7.416e+00 m^2/s^2, with low values indicating efficient flow.

4.6 AXIAL SPACING AXIAL VELOCITY PROFILE

4.6.1 Axial spacing axial velocity profile plot at 5000 rpm

The axial velocity profiles for the cases H0.5-V0.5, H1-V1 and H1.5-V1.5 are shown in Figure 4.95, 4.96 and 4.97. From these graphs, it can be clearly understood, at which configuration, where the flow reaches the maximum and highest peak. The red line indicates the In-Flow velocity. The blue line indicates the Out-Flow velocity and the green line is for Mid Flow velocity. Vortices and wakes are formed in this operation. Vortices are typically formed due to differences in pressure and velocity within a fluid. In the case of a rotor blade, vortices often form at the tips (tip vortices) due to the high-pressure difference between the upper and lower surfaces of the blade. Vortices can significantly affect the lift and drag of an aerodynamic surface. Tip vortices, for example, are associated with induced drag on rotor blades.

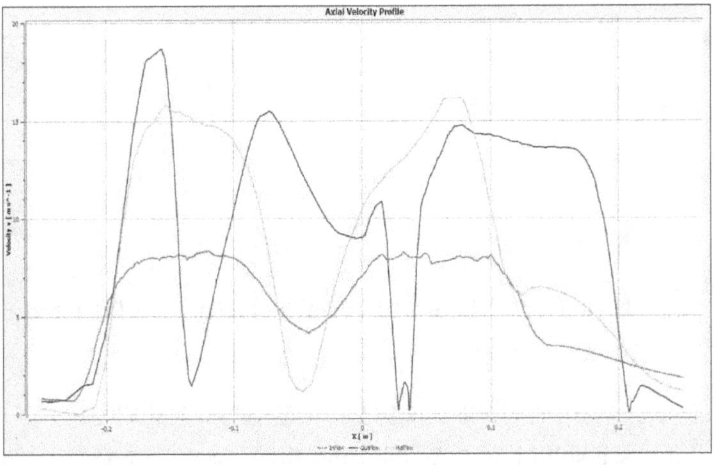

Figure 4.95 H0.5-V0.5 Axial Velocity Profile plot at 5000 rpm

In a rotorcraft, vortices from one blade can interact with subsequent blades, influencing overall performance.

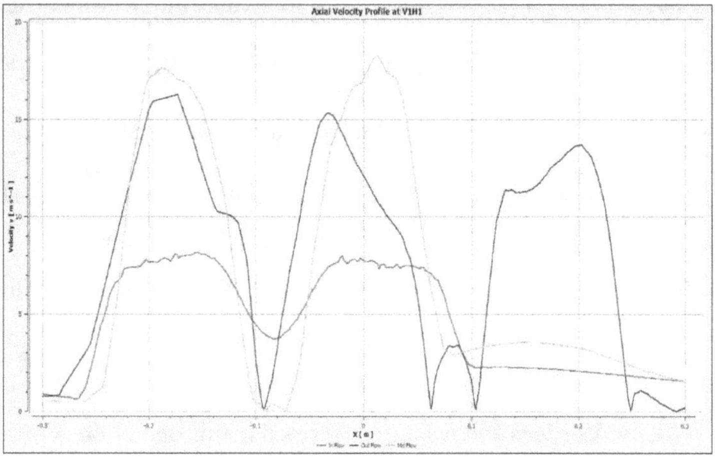

Figure 4.96 H1-V1 Axial Velocity Profile plot at 5000 rpm

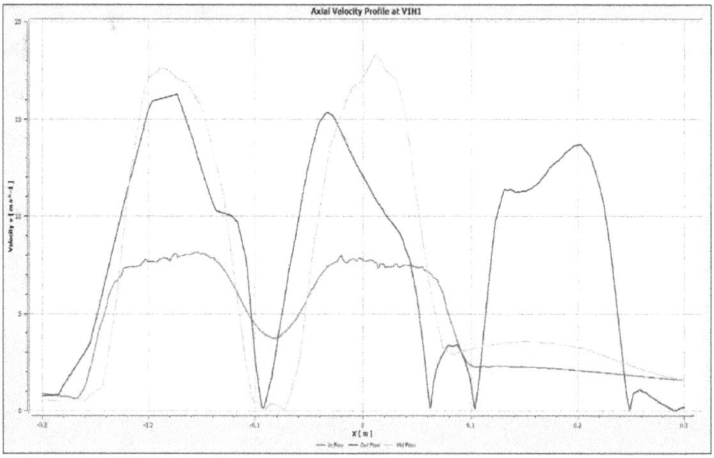

Figure 4.97 H1.5-V1.5 Axial Velocity Profile plot at 5000 rpm

Wakes form due to the obstruction and subsequent displacement of fluid by a solid body. As the fluid flows past the body, it separates from the surface and creates a region of lower velocity and higher turbulence downstream. Common examples of wakes include the turbulent flow behind a ship, the downwash behind a helicopter rotor, and the wake turbulence behind an aircraft. At the H0.5-V0.5, moderate wake interference is created due to closer spacing between blades. Limited space for vortices to develop fully. Baseline performance with moderate efficiency and stability. At the second case H1-V1, reduced wake interference with more space for wakes to dissipate and Better development of vortices, enhancing energy transfer are noticed. As such achieved the improved performance with higher efficiency and better stability. In the last case H1.5-V1.5, minimal wake interference is observed, allowing for very stable airflow. Optimal vortex development and controlled interaction helps in attaining maximum performance with the highest efficiency and stability. It can be concluded that larger distances allow wakes to stabilize and dissipate, minimizing their negative impact on subsequent blades. More space between blades permits better vortex formation and interaction, leading to more efficient energy transfer and reduced drag. In the H0.5-V0.5 configuration, the outflow reaches the maximum velocity 19m/s, and in the second and third cases, it reaches 16m/s as the maximum. The mid flow reaches maximum in the second and third cases with the values of 18.5m/s, and 17.5 m/s respectively. In the first case, the maximum velocity is 15.5m/s.

4.6.2 Axial spacing axial velocity profile plot at 8000 rpm

At 8000 rpm, the axial velocity profiles of the above-mentioned configurations are shown in Figures 4.98, 4.99, and 4.100. When rotors are close to each other as it is in the case of H0.5-V0.5, the vortices generated by one rotor can significantly interact with those from neighbouring or trailing rotors. This interaction can lead to increased drag, reduced efficiency, and complex aerodynamic behaviour. The vortices from one rotor may interfere with the performance of the adjacent rotor, causing turbulence and inefficiencies.

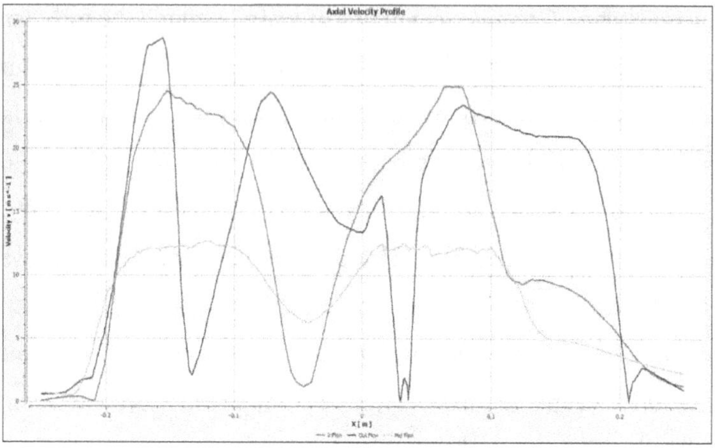

Figure 4.98 H0.5-V0.5 Axial Velocity Profile plot at 8000 rpm

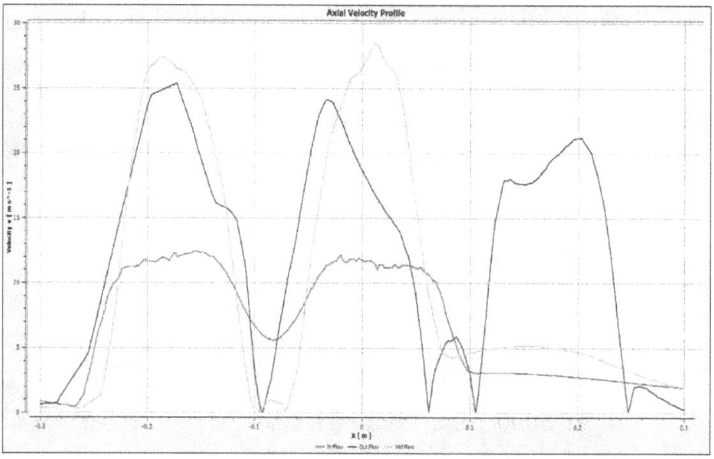

Figure 4.99 H1-V1 Axial Velocity Profile plot at 8000 rpm

As the distance between rotors increases, the vortices produced by each rotor have more space to dissipate before they can affect the neighbouring rotors. This separation reduces the likelihood of vortex interference and turbulence between rotors, leading to more stable and efficient operation. The rotors operate more independently, with less cross-meet in their aerodynamic effects. A similar pattern is observed at 8000 rpm as it is in the axial velocity profile at 5000 rpm but with intensified effect at all points. For example, at 5000 rpm, the configuration H1-V1 produced the maximum velocity of 17.5m/s while at 8000 rpm, it reached 28.5m/s. it shows the intensified flow with higher turbulence kinetic energy.

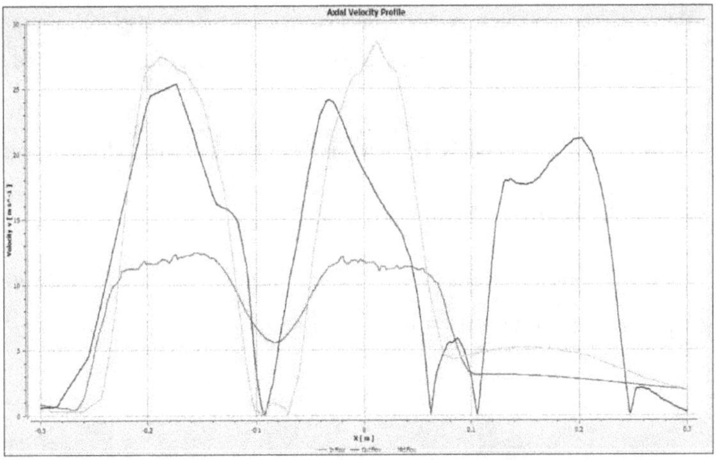

Figure 4.100 H1.5-V1.5 Axial Velocity Profile plot at 8000 rpm

4.7 MATHEMATICAL MODELLING RESULTS

4.7.1 Controller Tuning

The auto-tuning feature of Simulink can be used to manually or automatically adjust a PID controller. For every necessary input control, the PIDs are specifically adjusted. A simple approach can be used to manually tune PID controllers, albeit it may take some time and require iterations. Figure 4.101 illustrates how the P gain is adjusted initially and then progressively increased until it surpasses the necessary threshold when the gains of D and I are zero and this is the range of gain value. While tunning, the gains of D&I can be adjusted. The gain P is not calibrated in this setting. Once more, the visualization's scope is utilized to save time.

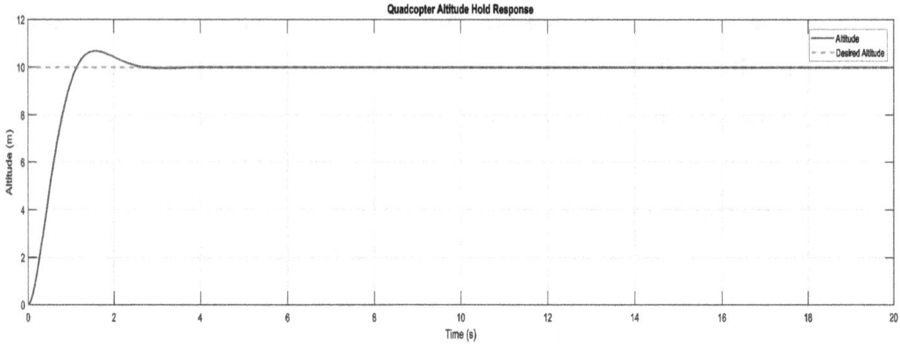

Figure 4.101 Altitude hold PID controller response

4.7.2 PID Tuning of Pitch and Roll

The roll and pitch of the PIDs follow the similar logic, hence their responses and level of gains are the same.

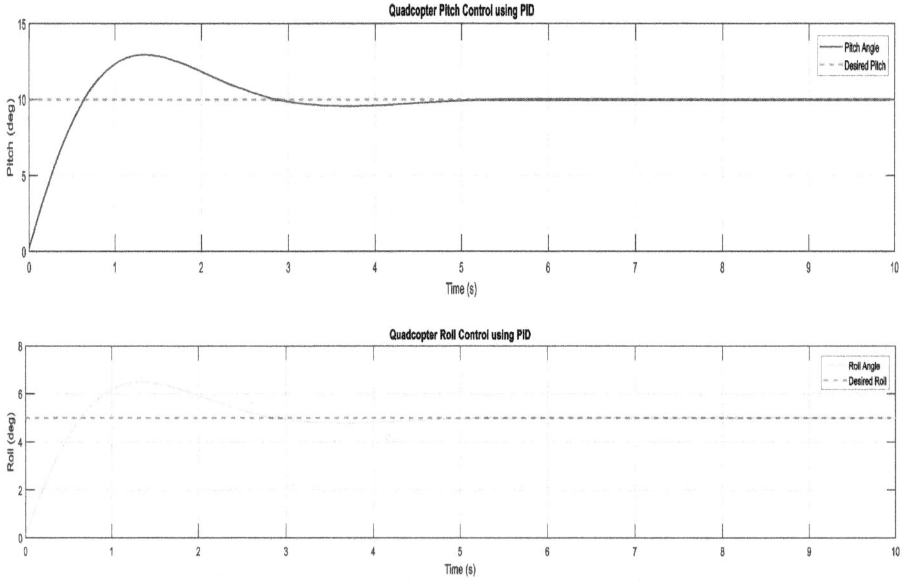

Figure 4.102 Pitch and roll response of the vehicle

The roll and pitch reaction of the vehicle when aiming for a 23-degree angle and trying to reach a 12-meter altitude at the same time is shown in Figure 4.102. The gains of PID controllers are 10.38, 8.4, and 0.081 respectively.

4.7.3 Yaw PID Tuning

In order to control yaw independently of roll and pitch, PID parameters must be adjusted. Because the yaw PID dependent on the four rotors torque, it is difficult to manage. Figure 4.103 shows the yaw controller's tuned response when is 12 degrees is the planned yaw rate and 1 is the vehicle altitude command. A tuned yaw PID controller gains are 4.63,2.36, and 0.03 respectively.

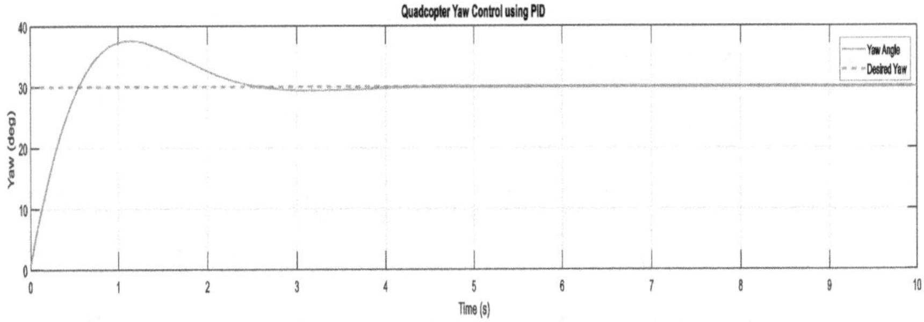

Figure 4.103. Yaw controller tuned response

4.7.4 Validation of results

The results are unusual since the PID controller's responsiveness and use in TAROT650 are being investigated for the first time. However, the modelling and experimental testing carried out on drones, particularly quadcopters, can be used to compare the pattern of the findings. In order to simulate and test the response of PID controller in a quadcopter MAV and with disturbance, Noordin et al. 2020 undertook simulation work. Figure 4.104 displays the results, which were cross-checked with one another. The TAROT 650 simulation was conducted using the identical protocol, incorporating wind as a disturbance, yielding comparable outcomes. Below are the reference findings for your comparison. The path of the followed trajectory was also computed among the altitude, pitch of the vehicle, then roll, and yaw.

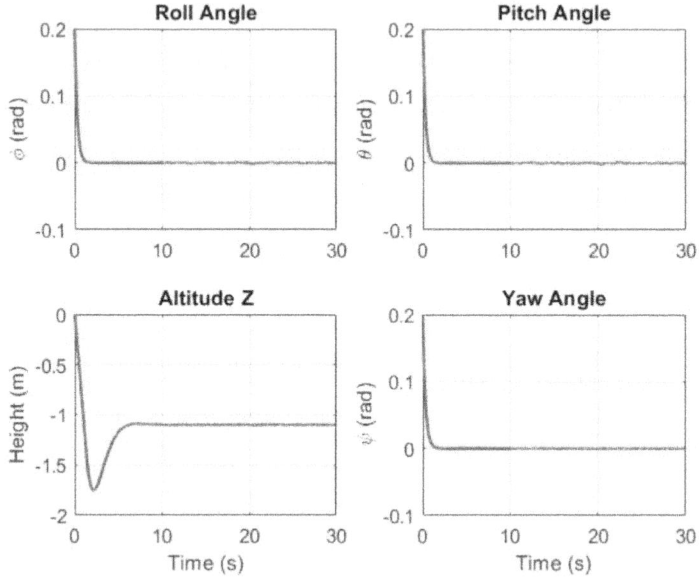

Figure 4.104 Roll, Pitch, Altitude and Yaw response of PID controller in Quadcopter MAV

4.7.5 Trajectory

A trajectory is designed in Simulink using the signal-builder block to demonstrate the vehicle's response to multiple signals given at once. The inertial trajectory is unpredictable because in the body coordinate system, the pitch, roll and yaw were controlled instead of the inertial coordinate system of X, Y, and Z. different simulated responses to different signal instructions. There are 120 seconds in the simulation. Figure 3.16 displays a snapshot of the control system that was utilized for simulating the quadcopter done by TAROT650.

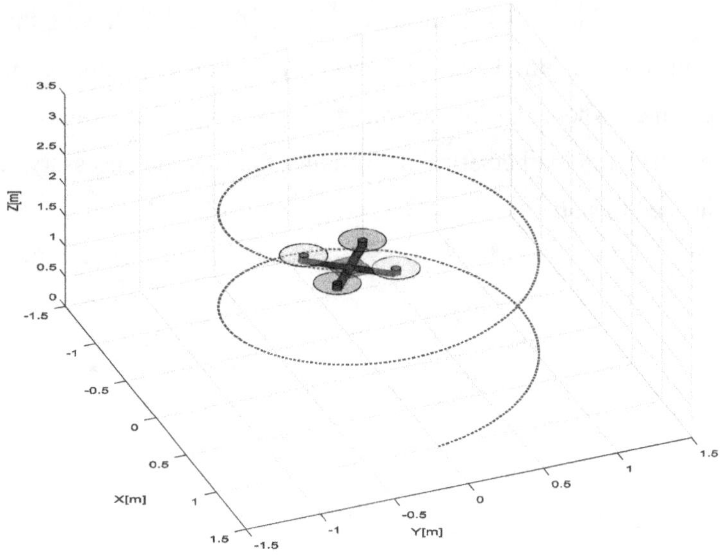

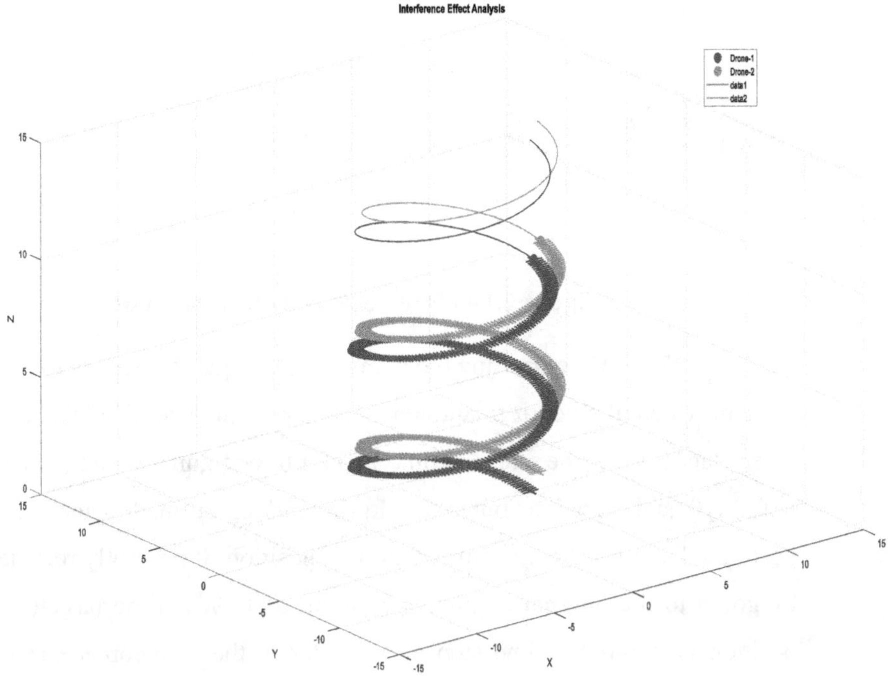

Figure 4.105 3D Space Trajectory-Drone-1(top), Drone-1 and Drone-2 (bottom)

Figure 4.105 shows the plot of drone-1 following the spiral (helical) trajectory in left, Right plot shows the simulated drones moving in closer distance where the interaction effects are optimum. Figure 4.106 shows the results from mathematical analysis for the specific necessary command that is needed to start.

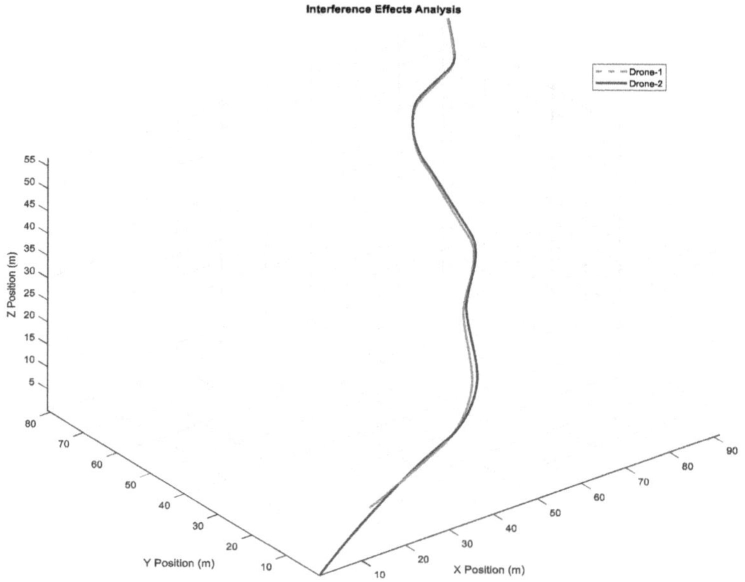

Figure 4.106 Interference Effect Analysis

The PID controlling block included the quadcopter's auto-correction system, which allowed it to automatically hover and stabilize. The correction takes place prior to the intended time step for three figures from first, according to the π, θ, and ψ position outcomes. In simulation, the quadcopter was initially positioned at 5 for the outcome of altitude position. It promptly rectified itself by going to the optimal beginning position of 1. When the targeted altitude position is 8 and the time step occurs after 6, the quadcopter can navigate swiftly to the target Z or altitude position.

CHAPTER 5

CONCLUSION

To maximize performance and stability, it is essential to recognize and measure the interference effects between UAV rotors. Enhancing flow characteristics and lowering drag can be achieved by varying the propeller spacing and keeping an appropriate distance between them. Finding this ideal distance is not only beneficial for the placement of rotors in single multirotor UAV but also in multirotor UAV transportation. Experimental, numerical and mathematical investigation are carried out for the selected 13inch EOLO 1345 propeller. ANSYS is used to numerically model different combinations with varying spacing and directions in order to analyse the mutual interaction between the rotors. For assessing the impact of the interference, the non-dimensional coefficients of thrust, torque and figure of merit (FM) are used.

The potential research findings obtained are mentioned below

1. Figure of Merit (FOM) is a rotor efficiency metric that effectively evaluates how near the rotor performs to its ideal performance. In lateral shift condition from LS 2.2 to LS 3, the Figure of Merit is increased. At lower lateral spacing, the reduced FM is observed due to the fact of Rotor Wake Interference, Increased Aerodynamic Interaction, Decreased Effective Disk Area and Mechanical and Acoustic Effects.

2. The interference between the wakes of neighbouring rotors grows as the lateral separation between them decreases. Every rotor creates a unique flow pattern that includes tip vortices, which are swirling of air created at the tips of the rotor blades, and downwash, which is air forced downward by the rotor. When rotors are placed closely together, the inflow conditions of nearby rotors may be impacted by the downwash of one rotor. For a given amount of thrust,

this disrupted input may result in less effective operation and higher power consumption.

3. At the LS 3, the optimum and completely the increased trend FM at all rpm is noticed due to the Enhanced downwash velocity and reduced tip vortex. At certain distance, interference between two rotors on a UAV or other multirotor aircraft can result in an increase in thrust that is positive. The primary cause of this phenomena is the positive interaction between the airflow fields produced by each rotor.

4. The presence of an adjacent rotor can interrupt the creation or persistence of tip vortices, which are a substantial source of energy loss in rotor systems. If the airflow from one rotor helps to disperse or weaken the tip vortices of another rotor more quickly, this can lower the induced drag and enhance the overall aerodynamic efficiency, resulting to a gain in force.

5. From LS 3.2 to LS 3.8, major fluctuation is noticed, that clearly shows that however the propellers are placed in long distance, the FM is reduced due to the flow separation, which is more unpredictable at this rotor spacing and the rotors' slightly increased contact with their own wake, and the inadequate thrust from the collapse of the suction forces on the tip. The increased variability to reduction in FM is indicative of this intensified interaction.

6. Axial spacing results are little different from lateral spacing. Each propeller is unequally affected by the rotor interference. At certain combination of axial and lateral spacing, the best performance is observed. H1.5-V1.5 shows better performance compared to others because of the less interference effects. At all rpm, the rising trend Figure of merit is noticed in H1-V1 where the interference effect is positively used. This is because of the rotor wake recovery. This is a scenario where the wake of one rotor, which typically contains rotational energy and reduced velocity (thus lower kinetic energy), is intercepted by

another rotor positioned strategically behind it. This second rotor can then harvest more kinetic energy from the wake, transforming some of the rotational energy back into useable thrust.

7. The quadcopter is stabilized using a PID controller, which was selected and successfully built using MATLAB. PID controller was used to test the quadcopter's operation and performance and produce the desired results. For the Simulink model of the quadcopter, a PID based controller is created and optimized, with a separate PID controller block for each variable that needed to be regulated. For visualization, a 3D trajectory animation is produced. This modelling and simulation help to understand the dynamics of the quadcopter and the parameter which affects the performance as well as the design of any quadcopter with X configuration.

www.ingramcontent.com/pod-product-compliance
Lightning Source LLC
LaVergne TN
LVHW010725061126
851857LV00014B/1349